MIX
Papier aus verantwortungsvollen Quellen
Paper from responsible sources
FSC® C105338

Kai Zhang

New Conjugated Polymers Based on Benzodifuranone and Diketopyrrolopyrrole

disserta
Verlag

Zhang, Kai: New Conjugated Polymers Based on Benzodifuranone and
Diketopyrrolopyrrole, Hamburg, disserta Verlag, 2012

ISBN: 978-3-95425-012-7
Druck: disserta Verlag, ein Imprint der Diplomica® Verlag GmbH, Hamburg, 2012

Bibliografische Information der Deutschen Nationalbibliothek
Die Deutsche Nationalbibliothek verzeichnet diese Publikation in der Deutschen
Nationalbibliografie; detaillierte bibliografische Daten sind im Internet über
http://dnb.d-nb.de abrufbar.

Die digitale Ausgabe (eBook-Ausgabe) dieses Titels trägt die ISBN 978-3-95425-013-4
und kann über den Handel oder den Verlag bezogen werden.

Universität zu Köln
Mathematisch-Naturwissenschaftliche Fakultät
Department für Chemie

Dieses Werk ist urheberrechtlich geschützt. Die dadurch begründeten Rechte,
insbesondere die der Übersetzung, des Nachdrucks, des Vortrags, der Entnahme von
Abbildungen und Tabellen, der Funksendung, der Mikroverfilmung oder der
Vervielfältigung auf anderen Wegen und der Speicherung in Datenverarbeitungsanlagen,
bleiben, auch bei nur auszugsweiser Verwertung, vorbehalten. Eine Vervielfältigung
dieses Werkes oder von Teilen dieses Werkes ist auch im Einzelfall nur in den Grenzen
der gesetzlichen Bestimmungen des Urheberrechtsgesetzes der Bundesrepublik
Deutschland in der jeweils geltenden Fassung zulässig. Sie ist grundsätzlich
vergütungspflichtig. Zuwiderhandlungen unterliegen den Strafbestimmungen des
Urheberrechtes.

Die Wiedergabe von Gebrauchsnamen, Handelsnamen, Warenbezeichnungen usw. in
diesem Werk berechtigt auch ohne besondere Kennzeichnung nicht zu der Annahme,
dass solche Namen im Sinne der Warenzeichen- und Markenschutz-Gesetzgebung als frei
zu betrachten wären und daher von jedermann benutzt werden dürften.

Die Informationen in diesem Werk wurden mit Sorgfalt erarbeitet. Dennoch können
Fehler nicht vollständig ausgeschlossen werden und der Verlag, die Autoren oder
Übersetzer übernehmen keine juristische Verantwortung oder irgendeine Haftung für evtl.
verbliebene fehlerhafte Angaben und deren Folgen.

© disserta Verlag, ein Imprint der Diplomica Verlag GmbH
http://www.disserta-verlag.de, Hamburg 2012
Hergestellt in Deutschland

New Conjugated Polymers

Based on Benzodifuranone and Diketopyrrolopyrrole

Inaugural-Dissertation

zur

Erlangung des Doktorgrades

der Mathematisch-Naturwissenschaftlichen Fakultät

der Universität zu Köln

vorgelegt von

Kai Zhang

aus V.R. China

Köln 2010

Berichterstatter:

Prof. Dr. Bernd Tieke

Prof. Dr. Ulrich K. Deiters

Einführung

Ziel dieser Arbeit war die Herstellung und Charakterisierung neuer konjugierten Polymere aud Basis von Diketopyrrolopyrrol (DPP) und Benzodifuranon (BZDF) als aktive Materialien für optoelektronische Anwendungen. Die Polymere wurden mit Hilfe der metal-katalysierten Kupplungen nach Suzuki, Stille, Sonogashira, und Yamamoto und durch anodische Elektropolymerization hergestellt. Die resultierenden Polymere wurden mit Hilfe der Gelpermeationschromatograpie, NMR-, UV/vis-, und Fluoreszenzspektroskopie, Massenspektrometrie sowie Zyklovoltammetrie untersucht.

Mit Hilfe der Elektropolymerization wurden Polymere auf Basis der 2,3,5,6-arylierten DPP-Einheit mit Comonomeren wie Thiophen, EDOT oder EDTT hergestellt. Sie weisen eine konjugierte Polymerkette auf, die durch die 3- und 6-Positionen der zentralen DPP-Einheit verläuft. Ihre HOMO-LUMO-Bandlücke liegt zwischen 1,3 und 2,0 eV und sie zeigen elektrochrome Eigenschaften. Polymere, deren Kette durch die 2- und 5-Positionen der DPP-Einheit verläuft, sind dagegen nicht konjugiert. Sie weisen eine Bandlücke von 3 eV auf. Die vernetzten DPP-Polymere zeigen sehr breite Absorptionsbanden mit Maxima um 500 nm und HOMO-LUMO-Bandlücken zwischen 1,3 und 2,0 eV. Die hergestellten Polymernetzwerke nach Yamamoto und Sonogashira weisen microporöse Eigenschaften auf. Sie zeigen BET-Oberflächen bis zu 500 m^2g^{-1} und Mikroporositäten von 17 bis 38 %. Außerdem wurde der Einfluß der thermischen Abspaltung der Boc-Gruppe auf die optischen und elektrochemischen Eigenschaften von ThienylDPP-Polymeren untersucht. Es zeigt sich, dass die Polymere nach der Abspaltung der Boc-Gruppe breitere Absorptionsbanden, höhere HOMO- und niedrigere LUMO-Levels aufweisen.

Desweiteren wurden konjugierte Polymere auf Basis von Benzodifuranon, die für die Anwendungen wie Photovoltaik oder elektrochrome Bauteile geeignet sein können, durch Palladium-katalysierte Suzuki-Kupplung oder anodische Elektropolymerisation hergestellt. Die Polymere zeigen sehr breite Absorptionsbanden mit Maxima bis zu 600 nm und Extinktionskoeffizienten von 25 000 bis 60 000 L $mol^{-1}cm^{-1}$. Außerdem weisen sie sehr kleine HOMO-LUMO-Bandlücken von 0,4 bis 1,2 eV auf. Oxidations- und Reduktionsprozesse sind reversibel. Reversible elektrochrome Eigenschaften der Polymere treten bei der kathodischen Reduktion auf. Die Farbe ändert sich von dunkelrot nach fast farblos.

Abstract

In this work synthesis and properties of new conjugated polymers containing diketopyrrolopyrrole (DPP) and benzodifuranone (BZDF) chromophores in the main chain are developed and investigated, which are suitable as active materials in optoelectronic applications. The polymers were synthesized by metal-mediated or -catalyzed Suzuki, Stille, Sonogashira and Yamamoto cross-coupling polycondensation reactions, or by using anodic electropolymerization. The resulting polymers were characterized using gel permeation chromatography, NMR-, UV/vis-, and fluorescence spectroscopy, mass spectroscopy, and cyclic voltrammetry.

Using anodic electropolymerization, polymers containing 2,3,5,6-arylated DPP were prepared. The incorporation of the tetra-functional DPP chromophore in a linear polymer chain via its 2- and 5- aryl units results in non-π-conjugated polymers exhibiting HOMO-LUMO band gaps of about 3 eV, whereas incorporation via the aryl units in the 3- and 6- positions results in π-conjugated polymers with HOMO-LUMO gaps between 1.3 and 2.0 eV, and the polymers exhibit electrochromic properties. The cross-linked polymers show a very broad absorption with maxima around 510 nm, exhibiting HOMO-LUMO band gaps between 1.3 and 2.0 eV. Cross-linked polymer networks prepared according to Yamamoto and Sonogashira coupling show microporous properties. They exhibit BET surface areas up to 500 $m^2 g^{-1}$, and show micoporosities in the range from 17 to 38 %. In addition, the influence of the thermal cleavage of the Boc-group on the optical and electrochemical properties of the Boc-protected ThienylDPP-polymers was studied. It was shown that, after removal of the Boc-groups, the polymers generally exhibit broader absorption bands and higher HOMO- and lower LUMO-levels, resulting in lower HOMO-LUMO gaps.

Furthermore, entirely new conjugated polymers containing the benzodifuranone (BZDF) chromophore in the main chain were synthesized. The polymers were prepared by palladium-catalyzed Suzuki cross-coupling polycondensation, or by anodic electropoly-merization, and are of potential interest for photovoltaic or electrochromic devices. The polymers show very broad absorption bands with maxima up to 600 nm, exhibiting extinction coefficients in the range from 25 000 to 60 000 L $mol^{-1} cm^{-1}$. The polymers exhibit very low band gaps in the range from 0.40 to 1.2 eV. Oxidation and reduction processes are reversible, which suggests a strong donor-acceptor character. Electrochromic properties were found upon reduction, the polymers showing colour changes from dark red to almost transparent.

Acknowledgement

First of all, I would like to express my deepest thanks to my doctoral supervisor Prof. Bernd Tieke for the opportunity to work on this fascinating topic in his group, and for his invaluable advice, guidance, inspiration, patience and support throughout my doctoral studies. With his comprehensive knowledge, enthusiasm, manner of scientific thinking, he has given me a role model as a researcher. I am very grateful that he allowed me to take part of international cooperations in Scotland.

I sincerely thank Dr. Raman Rabindranath and Dr. Yu Zhu for the helpful advice and support in my beginning time in the group. Especially all the evening discussions contributed largely to the accomplishment of this work.

My warmest thanks go to Prof. Peter Skabara of the Department of Pure and Applied Chemistry at University of Strathclyde, Glasgow, Scotland, for his guidance, encouragement, and hospitality in Glasgow. Working in his group was a great pleasure.

Dr. Filipe Vilela, Dr. John Forgie, and Dr. Alexander Kanibolotsky are thanked for the help and cooperation in Glasgow.

I thank my friend and also classmate Dr. Alexander Kühne for his help and advice during the Cologne and Glasgow time, and for introducing me to Scotland's local life.

I thank all the colleagues in my group, Anna, Ali, Ashraf, Belinda, Gülara, Iana, Irina, Julia, Kalie, Kristina, Monika, Philipp, Olga, Sogand, and Tatjana, for the pleasant and inspiring atmosphere in the group.

I also thank the students, Annett Rabis, Friederike Rohrbach, Malte Flory, Melanie Wienberg, Saman Ghasimi and Mirko Müller, whom I co-supervised during their laboratory internships, for their good performances and help.

Many thanks go to Mrs. Regina Ryba-Cheng and Mrs. Anna Maier for their help as secretaries.

I thank Prof. Ulrich K. Deiters for his attendance as a referee.

I thank Mr. James Morrow for performance of the gas sorption measurements, and Ms Ruth Brucker for the SEM measurements.

I acknowledge the financial support and supply on DPP pigments by BASF Schweiz AG, Basel, Switzerland, and Dr. Mathias Düggeli, Dr. Roman Lenz, and Dr. Pascal Hayoz, for their cooperation and helpful discussions. Mr. Mahmoud Zaher Eteish and Mr. Stefan Läuger are thanked for their help at BASF during the work in Basel, Switzerland.

I thank my friends Mirko and Hao for their invaluable advice, support and keeping up with my sometimes foolish moods in these years.

At last but not least, I would like to express my heartfelt thanks to my parents for their advice, encouragement, and unconditional love and support through these years.

TABLE OF CONTENS

	List of acronyms	IV
1	**Introduction**	**1**
2	**Overview**	**2**
2.1	Concept of the work	2
2.2	Diketopyrrolopyrrole	4
2.2.1	DPP monomers	4
2.2.2	DPP-containing polymers	7
2.3	Benzodifuranone	14
2.4	Metal-mediated and -catalyzed coupling reactions	17
2.5	Electropolymerization	22
2.6	Electronic applications of conjugated polymers	28
2.6.1	Organic photovoltaic cells	28
2.6.2	Electroluminescent applications	31
2.6.3	Electrochromism	33
2.7	Conjugated microporous polymers	34
3	**Conjugated polymers based on diketopyrrolopyrrole (DPP)**	**40**
3.1	2,5- and 3,6-Di-substituted pyrrolo[3,4-c]pyrrole-1,4(2H,5H)-dione (DPP)- based conjugated linear polymers	40
3.1.1	Synthesis and properties of monomers	41
3.1.2	Electropolymerization and properties of polymers	44
3.2	2,3,5,6-Tetra-substituted pyrrolo[3,4-c]pyrrole-1,4(2H,5H)-dione (DPP)- based conjugated polymers	53

3.2.1	2,3,5,6-Tetra-functionalized pyrrolo[3,4-c]pyrrole-1,4(2H,5H)-dione (DPP) monomers	53
3.2.2	Electropolymerization, optical and electrochemical properties of the polymers	59
3.3	Conjugated mesoporous poly(pyrrolo[3,4-c]pyrrole-1,4(2H,5H)-dione) (DPP) networks	66
3.4	Oligomers based on 2,3,5,6-tetraphenylpyrrolo[3,4-c]pyrrole-1,4(2H,5H)-dione (DPP)	75
3.5	Conjugated polymers based on thienyl-DPP	79
3.5.1	Conjugated alternating polymer containing Boc-substituted thienyl-DPP	79
3.5.2	Conjugated copolymers based on Boc-ThDPP	85
3.5.3	New thienyl-DPP	90
3.6	Conclusion	92
4	**Conjugated polymers based on benzodifuranone**	**94**
4.1	Symmetrical benzodifuranone-based conjugated polymers prepared via electropolymerization	94
4.1.1	3,7-Diphenylbenzo[1,2-b:4,5-b']difuran-2,6-dione-based conjugated polymers prepared via electropolymerization	96
4.1.1.1	Synthesis and properties of electropolymerizable monomers	96
4.1.1.2	Polymers prepared via electropolymerization	100
4.1.2	3,6-Diphenylbenzo-[1,2-b:6,5-b']difuran-2,7-dione-based conjugated polymers prepared via electropolymerization	106
4.1.2.1	Synthesis and properties of electropolymerizable monomers	108
4.1.2.2	Polymers prepared via electropolymerization	112
4.2	Symmetrical benzodifuranone-based conjugated polymers prepared via palladium-catalyzed cross-coupling polycondensation	118
4.2.1	Conjugated polymers prepared via Suzuki cross-coupling polycondensation	119
4.2.1.1	Synthesis and properties of the monomers	119

4.2.1.2	Synthesis and properties of the polymers	121
4.3	Symmetrical benzodifuranone-based conjugated polymers prepared via chemical oxidative polymerization	125
4.4	Unsymmetrical benzodifuranone-based conjugated polymers	129
4.4.1	Synthesis and properties of unsymmetrical benzodifuranone monomers	129
4.4.2	Polymers prepared via Suzuki and Stille cross-coupling polycondensations	132
4.4.3	Polymers prepared via electropolymerization	136
4.4.3.1	Synthesis and properties of the monomers	136
4.4.3.2	Polymers prepared via electropolymerization	139
4.5	Diphenylnaphthodifurandione	142
4.6	Conclusion	145
5	**Experimental part**	**144**
5.1	Materials	144
5.2	Instrumentation and general procedures	144
5.3	Synthesis	146
5.3.1	Diketopyrrolopyrrole-based conjugated polymers	146
5.3.2	Conjugated polymers based on benzodifuranone	172
5.3.2.1	Conjugated polymers based on symmetrical benzodifuranone	172
5.3.2.2	Conjugated polymers based on unsymmetrical benzodifuranone	188
6	References	204

List of acronyms

δ	Chemical shift
λ	Wavelength
Φ	Photoluminescence quantum yield
A	Ampere
BZDF	Benzodifuranone
cm	Centimeter
CV	Cyclic voltammertry
Da	Dalton
DCM	Dichloromethane
DMF	N,N-Dimethylformamide
DMSO	Dimethylsulfoxide
DPP	Diketopyrrolopyrrole
EC	Electrochromic
EDOT	3,4-Ethylenedioxithiophene
EDTT	3,4-Ethylenedithiathiophene
EL	Electroluminescence
FL	Fluorescence
g	Gram
GPC	Gel permeation chromatography
h	Hour
HOMO	Highest occupied molecular orbital
IR	Infrared
LED	Light emitting diode
LUMO	Lowest unoccupied molecular orbital
min	Minute
mm	Milimeter
nm	Nanometer
Mp	Melting point
NMR	Nuclear magnetic resonance
PD	Polydispersity
PFET	Polymeric field effect transistors
PL	Photoluminescence
PLED	Polymeric light emitting diodes
PVC	Polymeric photovoltaic cells

RT	Room temperature
THF	Tetrahydrofuran
UV	Ultraviolet
V	Volt
vis	Visible light
wt.%	Weight percent

1 Introduction

The tremendously growing demand for energy throughout the world has addressed great emphasis on exploring clean and renewable energy resources. One of the most important long-term solutions is to harvest energy directly from the sunlight using photovoltaic technology. Since it was discovered in 1970 that polyacetylene can be rendered electrochemically conductive,[1-3] polymeric photovoltaic cells (PVC)[4-7] have become a promising alternative due to the easy fabrication onto large areas of flexible substrates and the possibility of solution processing at low cost.[8, 9] Due to their outstanding optical and electrochemical properties, conjugated polymers also have attracted much attention in other application fields such as polymeric light emitting diodes (PLED)[10] and polymeric field effect transistors (PFET).[11] Among the applications, high grade light emitting diodes with very high efficiencies and considerably long life times are already commercially available. For the pioneering work in the field, A. MacDiarmid, A. Heeger, and H. Shirakawa were awarded with the Nobel Prize in 2000. Since that time, a vast number of research works focusing on the preparation and modulation of the properties of conjugated polymers has been published.

In order to obtain polymeric electronic materials, it is necessary to design and synthesize conjugated polymers with suitable properties, such as low HOMO-LUMO gaps, broad absorption range, high electron and hole mobility, and appropriate molecular energy levels. One useful strategy to design new conjugated polymers for electronic applications is to incorporate chromophores which are highly absorbing or emitting in the visible and near infrared region into π-conjugated polymer chains. Potentially useful chromophores can be found among the various organic colourants, especially in the field of so-called "high-performance pigments" developed in the last three decades. [12] This work is concerned with the incorporation of such chromophores in conjugated polymer chains and the characterization of the optical and electrochemical properties of the resultant polymers.

2 Overview

In this chapter, the concept of this work and a brief overview on preparation, key properties and application fields of the conjugated polymers will be given.

2.1 Concept of the work

Among the so-called "high-performance pigments" 2,5-diketopyrrolo[3,4-c]pyrrole (DPP) and its derivatives have been one of the key subjects in our research group in the last decade. In our previous work,[13] as shown in Scheme 2.1, 3,6-diphenylfuro[3,4-c]furan-1,4-dione (DFF) was used as the starting compound for preparation of two isomeric bifunctional monomers **DPP1** and **2**. Corresponding linear polymers were synthesized and characterized. Continuing this subject, tetra-functional DPP monomers (*t*-DPP) will be synthesized. Polymerization will result in cross-linked polymers, which are suitable for electronic applications or for gas storage due to the porous surface area, will be obtained via metal-mediated and -catalyzed cross-coupling polycondensation reactions, or by using the method of electropolymerization under anodic conditions. For comparison, DPP-based polymers will also be synthesized under similar reaction conditions, and the influence of the substitution pattern on optical and electrochemical properties will be studied.

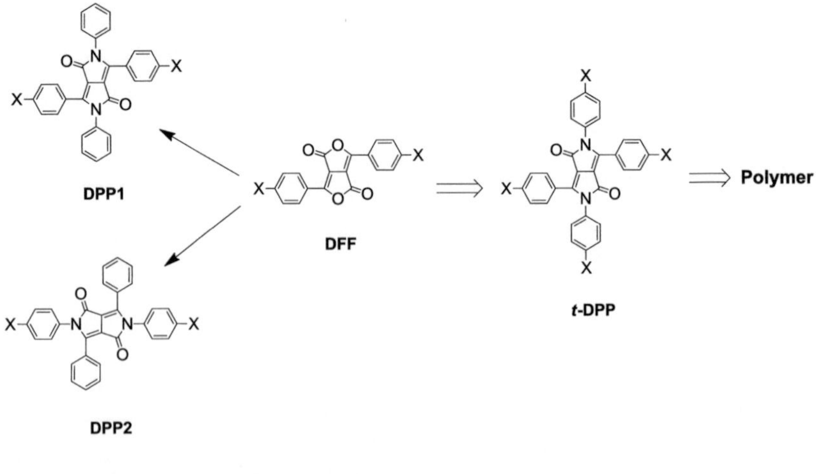

Scheme 2.1. Synthetic strategy for new DPP-based conjugated polymers.

In addition, thienylDPP monomers will be substituted with *t*-butoxycarbonyl (Boc)-group and the corresponding polymers will be synthesized via palladium-catalyzed Suzuki cross-coupling polycondensation reaction. A special emphasis will be paid to the influence of the thermal cleavage of the Boc-group on the optical and electrochemical properties of the polymers.

As a second topic, an entirely new conjugated polymer will be synthesized containing the benzodifuranone (BZDF) chromophore in the main chain. The BZDF moiety is related to the DPP chromophore mentioned above in the following way. Derived from the structure of 3,6-diphenylfuro[3,4-c]furan-1,4-dione (DFF), 3,6-diphenylfuro-[3,2-b]furan-2,5-dione (*iso*DFF) represents an isomer of DFF. Incorporation of a phenyl unit between the two furanone groups results in a molecule with a larger conjugated system, the compound being denoted as 3,7-diphenylbenzo[1,2-c:4,5-c']difuran-1,5-dione (BZDF) (Scheme 2.2).

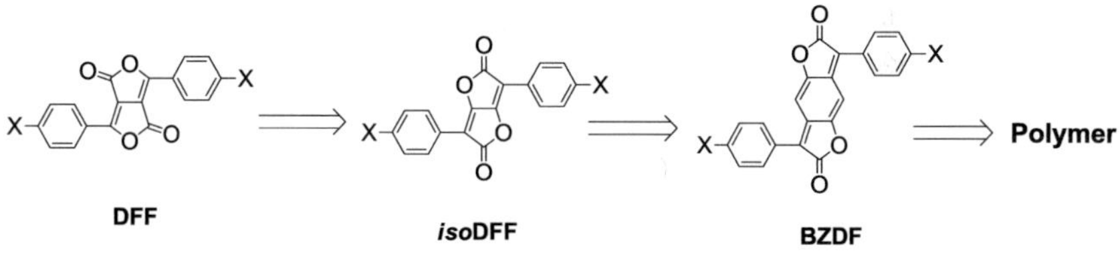

Scheme 2.2. Relation between structures of DFF and BZDF.

BZDF has been commercialized as disperse dye for textiles. Despite the early discovery of the benzodifuranone chromophores, only few research articles have been published. Incorporation of benzodifuranone chromophores into polymer chains has never been reported, although it might be interesting for its extended π-system and its quinoid structure, from which a strong light absorption in the visible can be expected. In order to realize BZDF-based polymers, the following work will be carried out:

- A synthetic method for preparation of a bifunctional benzodifuranone monomer suitable for palladium-catalyzed polycondensation reactions will be developed.
- Benzodifuranone monomers will be incorporated in conjugated polymer chains. The benzodifuranone-based conjugated polymers will be synthesized via palladium-catalyzed cross-coupling polycondensation reactions, or by using the electropolymerization method.

- Various comonomers containing electron-donating or -withdrawing groups will be used. The influence of the BZDF units on the optical and electrochemical properties of the polymers will be studied. New polymers exhibiting high HOMO- and low LUMO-level resulting in low band gaps, which are suitable for polymer solar cells, will be developed.
- The ease of reduction of polymers based on benzodifuranone will be used to prepare electrochromic devices and study their cathodic reduction using spectroscopic methods.

Furthermore, the polymers will be characterized using NMR-, UV/vis absorptions-, photoluminescence, and infrared spectroscopy, cyclic voltammetry, and gel permeation chromatography (GPC). The cross-linked polymer networks will be investigated using scanning electron microscope (SEM) and nitrogen adsorption and desorption analysis.

To better understand the research work, an overview of diketopyrrolopyrrole and its polymers, the benzodifuranone chromophore, metal-catalyzed cross-coupling reactions, and electrochemical reactions for the synthesis of the polymers, and application of the polymers in solar cells, electroluminescent and electrochromic devices, and as rigid microporous systems for gas storage will be given in the following.

2.2 Diketopyrrolopyrrole

2.2.1 DPP monomers

2,5-Diketopyrrolo[3,4-c]pyrrole (DPP) (Fig. 2.1) and its derivatives have been commercialized since the 1980s, and have been the subject of many patents, despite the fact that for a considerable time only a few publications dealt with these compounds. In recent years, a growing number of polymer chemists and physicists have become interested in DPPs since it was shown that DPP-containing polymers exhibit light-emitting and photovoltaic properties.

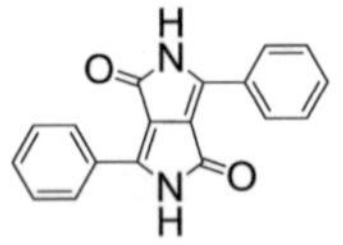

Fig. 2.1. Structure of 3,6-diphenyl-substituted 2,5-diketopyrrolo[3,4-c]pyrrole (DPP).

3,6-Diphenyl-substituted DPP (diphenylDPP) was first reported by Farnum et al. in 1974 (Scheme 2.3).[14] Instead of the desired product, diphenylDPP was obtained in low yield.

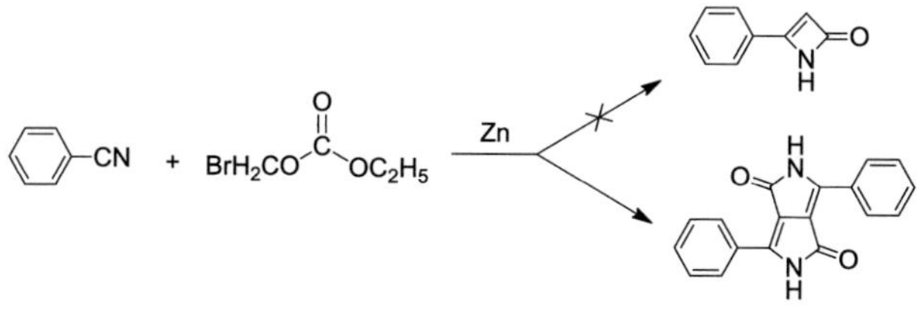

Scheme 2.3. Synthetic route to DPP according to Farnum et al.

In 1983, Iqbal, Cassar, and Rochat[15, 16] reported an elegant synthetic pathway for DPP derivatives. After the reaction of benzonitrile (or other aromatic nitriles) with succinic acid diesters, DPP derivatives could be prepared in a single reaction step in high yield (Scheme 2.4). Numerous DPP derivatives were synthesized since, their colour ranging from orange yellow via red to purple. Many DPP derivatives exhibit a high photostability in the solid state, weather fastness, deep colour, luminescence with large Stokes-shift, and a brilliant red colour enabling technical applications in colouring of fibers, plastics, surface coatings such as prints or inks.

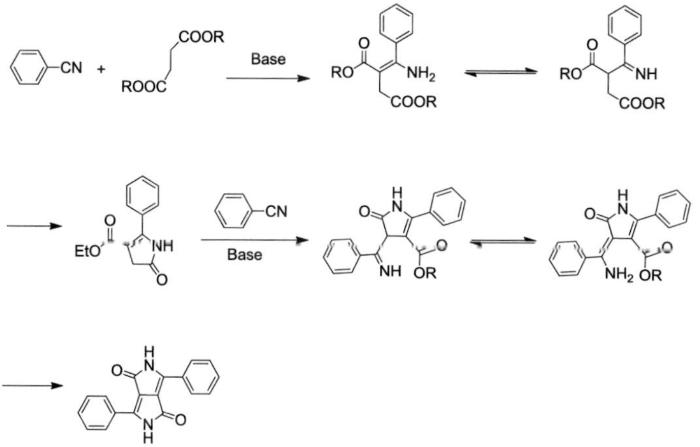

Scheme. 2.4. Synthetic route to DPP using aromatic nitriles and succinic esters.

The electron-withdrawing effect of the lactam units supplies the chromophore with a high electron affinity. Strong hydrogen bonding between the lactam units favours the chromophores into physically cross-linked chain structures in the solid state, which is the origin for the poor solubility (Fig. 2.2).[17, 18] Short distances between the chromophore planes (0.336 nm) and phenyl ring planes (0.354 nm) enable π–π-interactions via molecular orbital overlapping and exciton coupling effects.[17-19] Electronic interactions and strong intermolecular forces also cause a high thermal stability up to 500 °C.

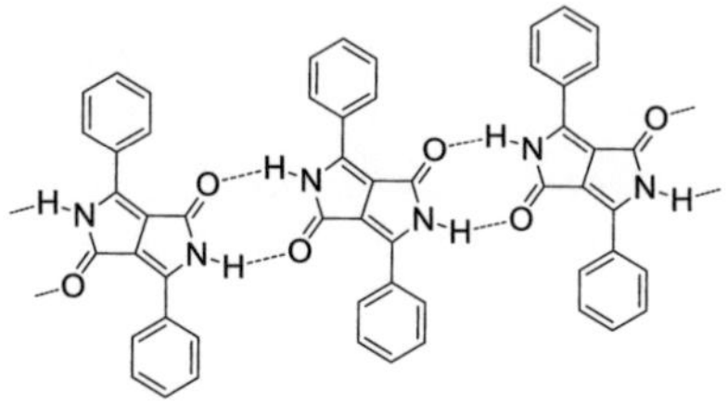

Fig. 2.2. Hydrogen bonds between the DPP units in the solid state.

For chemical incorporation into conjugated polymers, the solubility of the DPP compound needs to be increased, and the chromophore needs to be functionalized with polymerizable groups. The solubility can be increased by N-alkylation[20], arylation[21] or acylation[22] of the lactam units preventing hydrogen bond formation between the chromophores. Polymerizable groups can be attached to the aryl units in the 3- and 6-position of the central DPP chromophore,[23] or to the lactam substituent groups.[24, 25] Suitable polymerizable groups are halogen atoms (especially bromine and iodine), hydroxyl, trifluoromethylsulfonate, or aldehyde groups (Fig. 2.3).

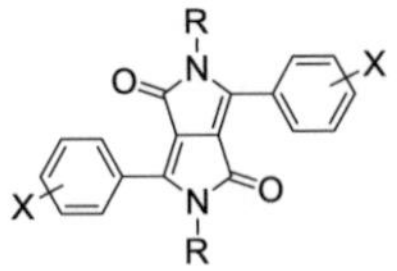

R = Alkyl, Aryl, -CH$_2$COOC$_2$H$_5$, -CH$_2$CH=CH$_2$, -CH$_2$CN

X = -SO$_3$H, -Cl, -CN, -Br, -CH$_3$

Fig. 2.3. Various DPP derivatives.

2.2.2 DPP-containing polymers

The first DPP-based polymer was described by Chan et al. in 1993.[26] Conjugated block copolymers containing phenylene, thienylene and N-alkyl substituted diphenylDPP units in the main chain were synthesized by Stille coupling (Fig. 2.4). Photorefractive polymers were prepared containing a conjugated main chain and nonlinear optically active (nlo) chromophores in the side chain. DPP was incorporated in the polymers as a sensitizer for charge carrier generation.

Some years later, Eldin and coworkers described DPP-containing polymers obtained by radical polymerization of bis-acryloyl-substituted DPP derivatives.[24, 25] Polymer networks containing non-conjugated, copolymerized DPP units were prepared. Linear DPP-containing polyesters and polyurethanes were first described by Lange and Tieke in 1999.[27] The polymers were soluble and could be cast into orange films exhibiting a strong fluorescence with maxima at 520 nm and a large Stokes shift of 50 nm. Due to the aliphatic structure of the main chain, the thermal stability was rather poor. Photoluminescent polyelectrolyte-surfactant complexes were obtained from an amphiphilic, unsymmetrically substituted DPP-derivative by complex formation with polyallylamine hydrochloride or polyethyleneimine.[28] The complexes exhibit a mesomorphous structure, the glass transition depending on the structure of the polyelectrolyte.

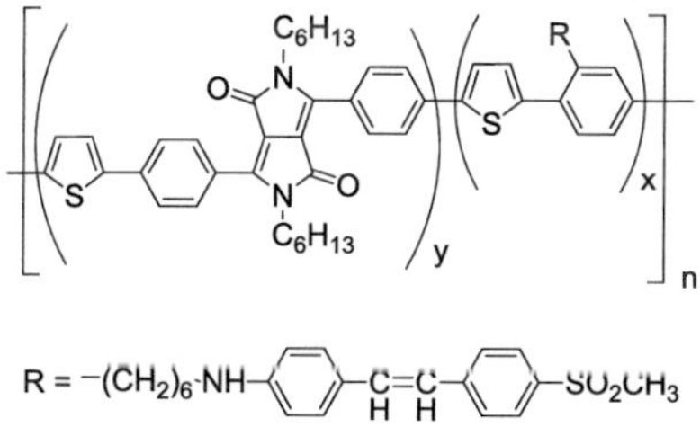

Fig. 2.4. The first reported DPP containing conjugated polymer.

The first synthesis of conjugated DPP-polymers and copolymers by Pd-catalyzed Suzuki coupling was reported by Tieke and Beyerlein in 2000 (Fig. 2.5).[29] The polymers contained N-hexyl-substituted diphenylDPP units and hexyl-substituted 1,4-phenylene units in the main chain. Molecular weights up to 21 kDa were determined. Compared with the monomer, the optical absorption of the polymer in solution was bathochromically shifted by 12 nm, the maximum appearing at 488 nm. The polymer showed a bright red fluorescence with a maximum at 544 nm. In addition to the alternating copolymer, also copolymers with lower DPP content were prepared. All copolymers showed the DPP absorption at 488 nm, the ε-value being a linear function of the DPP content. By UV irradiation the copolymers gradually decomposed. The rate of photodecomposition increased with decreasing DPP:phenylene comonomer ratio. Two different photoprocesses were recognized: a slow process originating from the absorption of visible light by the DPP chromophore, and a rapid one originating from additional absorption of UV-light by the phenylene comonomer unit, followed by energy transfer to the DPP chromophore. The actual mechanism of photodecomposition remained unclear. Comparative studies indicated that conjugated DPP-containing polymers are considerably more stable than the DPP monomers or non-conjugated DPP-polymers.

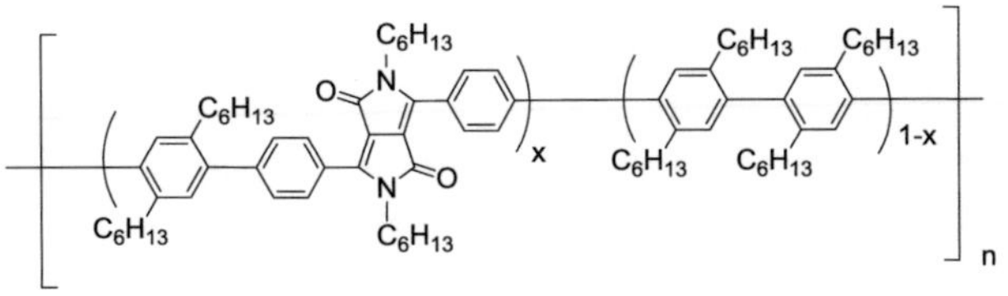

Fig. 2.5. Conjugated DPP-containing polymers prepared by Pd-catalyzed Suzuki coupling.

Smet et al. used a step sequence of Suzuki couplings to prepare rod-like DPP-phenylene oligomers with well-defined length.[30] The resulting oligomers contained three, five and seven DPP units, respectively. Unfortunately, the effect of the chain length on absorption and emission behavior was not reported. A study on thermomesogenic polysiloxanes containing DPP units in the main chain was published in 2001.[31] Investigations of the thermotropic phase behaviour using polarizing microscopy revealed nematic and smectic enantiotropic phases. In the same year, a first study on electroluminescent (EL) properties of a DPP-containing conjugated polymer was reported. Beyerlein et al.[32] studied a DPP-dialkoxyphenylene copolymer in a multilayer device of ITO/DPP-polymer/OXD7/Ca/-Mg:Al:Zn and observed a red emission with maximum at about 640 nm. DeSchryver et al. synthesized dendrimer macromolecules with a DPP core.[33] Embedded in a coated polystyrene film, single dendrimer molecules could be imaged with a confocal microscope utilizing the strong fluorescence of the DPP core. It could be shown that the orientation of the absorption transition dipole of single dendrimer molecules in the film changed in a time span of seconds.

In recent years a number of studies were reported on synthesis, optical, electrochemical, and electroluminescent properties of conjugated DPP polymers. The polymers were prepared by Suzuki, Heck, and Stille coupling and other catalytic polycondensation reactions.

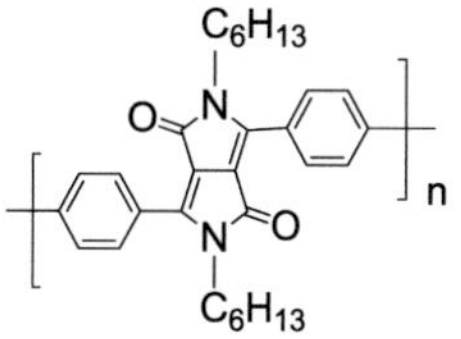

Fig. 2.6. DPP-containing polymer according to Rabindranath et al.

Rabindranath et al.[34] synthesized a new DPP polymer entirely consisting of aryl-aryl coupled diphenyl-DPP units (Fig. 2.6). In solution the polymer exhibits a bordeaux-red colour with absorption maxima of about 525 nm, and a purple luminescence with maxima around 630 nm, the Stokes-shift being about 105 nm. Cyclovoltammetric studies indicated quasi-reversible oxidation and reduction behaviour, the band gap being about 2 eV.

In a comprehensive study, Zhu et al. prepared a number of highly luminescent DPP-based conjugated polymers (Fig. 2.7).[35] The polymers consisted of dialkylated DPP units and carbazole, triphenylamine, benzo[2,1,3]thiadiazole, anthracene, or fluorene units in alternating fashion exhibiting yellow to red absorption and emission colours, and fluorescence quantum yields up to 86%. EL devices prepared with the polymer containing DPP and fluorene exhibited an external quantum efficiency (EQE) of 0.5% and a brightness at 20 V of 50 cd m^{-2} without much optimization. The maximum emission was at 600 nm, the turn-on voltage was 3.5 V. Cao et al.[36] prepared DPP-fluorene copolymers, the DPP content varying between 0.1 and 50%. It was found that absorption and emission spectra, both in solution and thin film, varied regularly with the DPP content in the copolymers. By an increase of the DPP content, the absorption only shifted by a few nanometers to longer wavelength, whereas the emission bathochromically shifted by more than 40 nanometers. EL properties of the copolymers were also studied. With increasing DPP content the EL colours varied from orange to red corresponding to CIE coordinates from (0.52, 0.46) to (0.62, 0.37). The best performance was achieved for an orange emitting device with a copolymer containing only 1% DPP units. The EQE was 0.45%, the maximum brightness 520 cd m^{-2}. At high DPP content, the EQE was lowered to 0.14%, and the brightness to 127 cd m^{-2}, similar to the results reported by Zhu et al.[35]

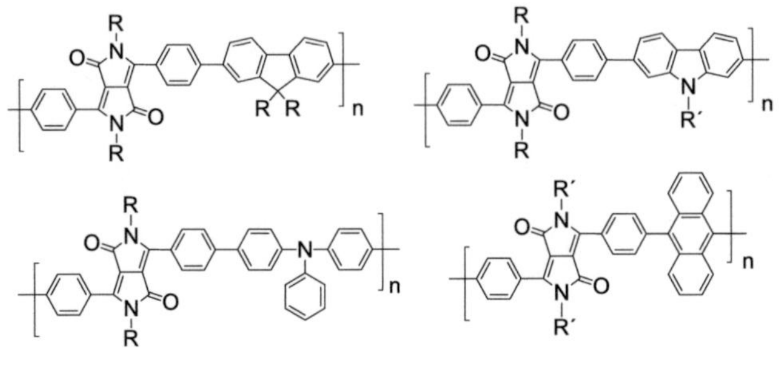

R = Hexyl, R´ = 2-Ethylhexyl

Fig. 2.7. Examples of DPP-containing polymers according Zhu et al.

Novel vinylether-functionalized polyfluorenes for active incorporation in common photoresist materials were described by Kühne et al.[37] Among the polymers investigated was a diphenylDPP-fluorene copolymer, the fluorene units carrying ethyl vinylether groups in the 9,9´-position. The vinyl ether functionality allowed for active incorporation of the light emitting polymers into standard vinyl ether or glycidyl ether photoresist materials, the polymers retaining their solution fluorescence characteristics. This enabled photopatterning of light-emitting structures for application in UV-down-conversion, waveguiding, and laser media.

Using Stille coupling, Zhu et al.[38] first succeeded in the synthesis of copolymers containing diphenylDPP and thiophene, bisthiophene, or 3,4-ethylenedioxythiophene (EDOT) units in alternating fashion, which exhibited a maximum absorption of 560 nm. A solution-cast film of the same polymer had a λ_{max}-value of 581 nm, and band gaps up to 1.7 eV, which were, due to the strong donor-acceptor interaction between the thiophene and the DPP units, considerabely smaller than for the previously reported DPP-based polymers (Fig. 2.8).

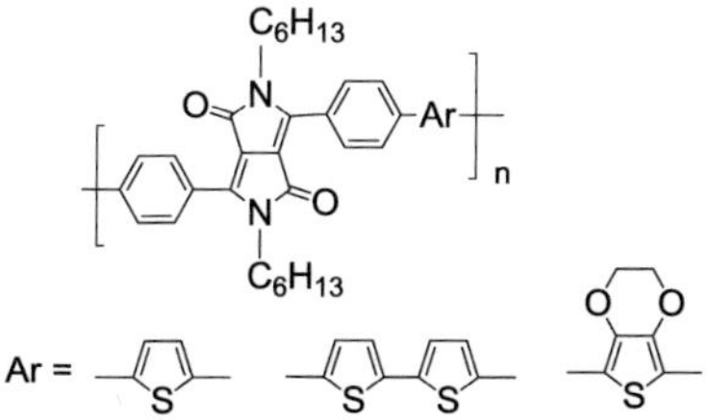

Fig. 2.8. DPP-containing polymers prepared by Stille-coupling.

In a further study[39] the incorporation of arylamine units in the main chain was attempted according to the synthetic pathway reported by Hartwig,[40-42] Buchwald,[43-45] and Kanbara.[46-49] The examples of the polymers are displayed in Fig. 2.9. Due to presence of electron-rich nitrogen atoms, the donor-acceptor interactions along the main chain were enhanced causing a red-shift of the absorption and emission. Furthermore, the presence of easily oxidizable nitrogen in the main chain gave rise to a lower oxidation potential of the polymers. The solutions of the polymers in chloroform exhibit a purple red colour with absorption maxima between 530 and 550 nm, and emission maxima from 610 to 630 nm. Fluorescence quantum yields are moderate (20 to 60%). The nitrogen atoms in the backbone lower the band gap of the polymers to approximately 1.9 eV.

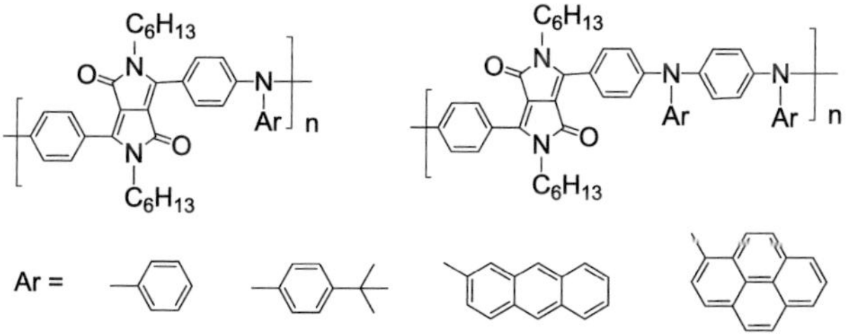

Fig. 2.9. DPP-containing polymers prepared by Buchwald-Hartwig coupling.

Interesting N-arylated diphenylDPP derivatives (also denoted as 2,3,5,6-tetraarylated DPP derivatives) were prepared by Zhang and Tieke.[13] Polymers containing the two isomeric DPP monomers were synthesized using Pd-catalyzed Suzuki coupling polycondensation reaction with fluorene as the comonomer unit (Fig. 2.10). While the properties of the two monomers are very similar, the optical and electrochemical properties of the two isomeric polymers are quite different. Suzuki coupling of **BrDPP1** and a fluorene diboronester derivative results in polymer **P-1** with fully conjugated main chain, the absorption being shifted by 15-25 nm compared with the monomer (Fig. 2.10). The same coupling reaction of **BrDPP2** results in polymer **P-2**, its π-conjugation being interrupted at the N-lactam units. Consequently, absorption and emission behaviour are not much different from the corresponding monomer, the band gaps of the two isomers being 2 and 2.3 eV, respectively. Absorption and emission colours are shown in Fig. 2.10.

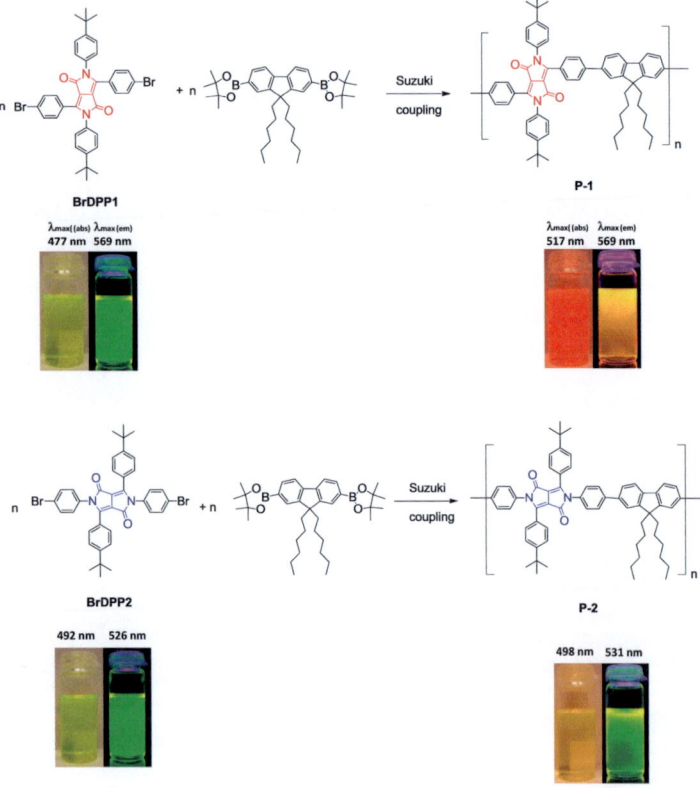

Fig. 2.10. Optical properties of DPP-containing copolymers based on two isomeric diphenylDPP monomer units.

ThiophenylDPP-based copolymers

The replacement of the phenyl groups in 3,6-diphenyl-substituted DPP derivatives for thiophenyl groups results in 3,6-(2-thiophenyl)-substituted DPP derivatives (thiophenylDPPs) with absorption maxima at about 530 nm, i.e., more than 50 nm bathochromically shifted compared to diphenylDPP. Corresponding comonomer and polymer structures are listed in Fig. 2.11.

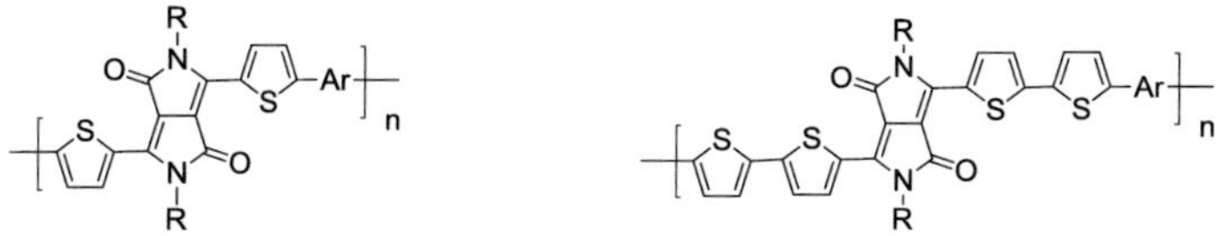

Important comonomers Ar

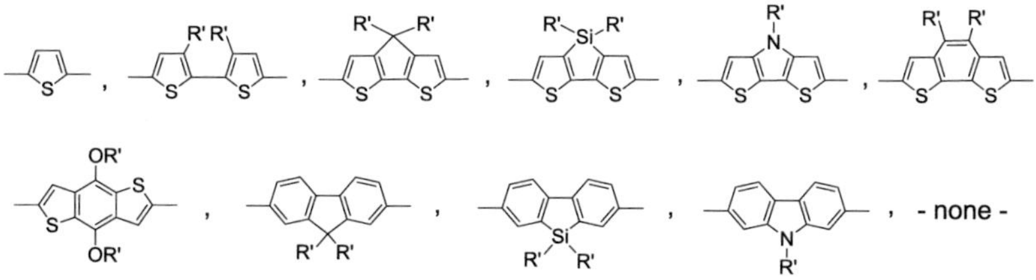

Typical alkyl groups R, R´:

n-hexyl, n-octyl, 2-ethylhexyl, n-dodecyl, 2-butyloctyl, 2-hexyldecyl, 1-octylnonyl, 1-decylundecyl

Fig. 2.11: Thiophenyl-DPP-based polymers.

Conjugated polymers containing thiophenylDPP in the main chain exhibit absorption maxima between 600 and 900 nm. Because of their small band gaps and high charge carrier mobilities, the polymers are interesting for applications in field effect transistors (FETs) and organic photovoltaic cells. Using Yamamoto coupling of a dibrominated thiophenylDPP derivative, Winnewisser et al.[50] succeeded in the preparation of the thiophenylDPP-based polymer, poly[3,6-bis(4´-dodecyl-[2,2´]bithiophenyl)DPP]. Using this material, an ambipolar near-infrared light-emitting transistor (LET) could be prepared. The LET exhibited hole and electron mobilities of 0.1 $cm^2 V^{-1} s^{-1}$ and up to 0.09 $cm^2 V^{-1} s^{-1}$, respectively. These values were higher than any other ones reported for solution-processed ambipolar transistors before. Janssen et al.[51] demonstrated the utility of thiophenylDPP-containing conjugated polymers for application in photovoltaic devices. Using a mixture of C_{70}PCBM and thiophenylDPP-based polymer as active layer, solar cells exhibiting a power conversion efficiency up to 4.0% could be fabricated. The polymer exhibits a band gap of 1.4 eV, the maximum is shifted to 810 nm indicating chain aggregation and ordering, which is an important prerequisite for the preparation of films with good photovoltaic performance. In subsequent studies the efficiency could be further increased, e.g. by preparation of so-called 'polymer tandem solar cells' consisting of two subcells converting different parts of the solar spectrum.[52] For such a cell, an efficiency of 4.9% could be reached. Encouraged by the good performance of thiophenylDPP-based solar cells, further polymers were recently synthesized and tested on their photovoltaic properties. Among them are alternating copolymers containing the thiophenylDPP, or bithiophenylDPP unit,[53] and carbazole,[53-55] fluorene,[53, 55-57] dibenzosilole, dithienosilole,[58] benzo[1,2-b;3,4-b]dithiophene,[59] benzo[2,1-b;3,4-b´]dithiophene,[56] dithieno-[3,2-b;2´,3´-d]pyrrole[55, 60] and cyclopenta[2,1-b;3,4-b´]-dithiophene[59] as comonomer units. Some of the polymers are suited for preparation of highly efficient polymer solar cells,[55] some also turned out to exhibit ambipolar charge transport[60] with hole and electron transport mobilities up to 0.04 $cm^2 V^{-1} s^{-1}$ and 0.01 $cm^2 V^{-1} s^{-1}$, respectively.[51]

2.3 Benzodifuranone

In 1960, Junek reported a red product, which was obtained by the reaction of benzoquinone with cyanoacetic acid without addition of any catalyst.[61] He predicted that a linear pentacenequinone was obtained. The synthetic pathway is displayed in (Scheme 2.5).

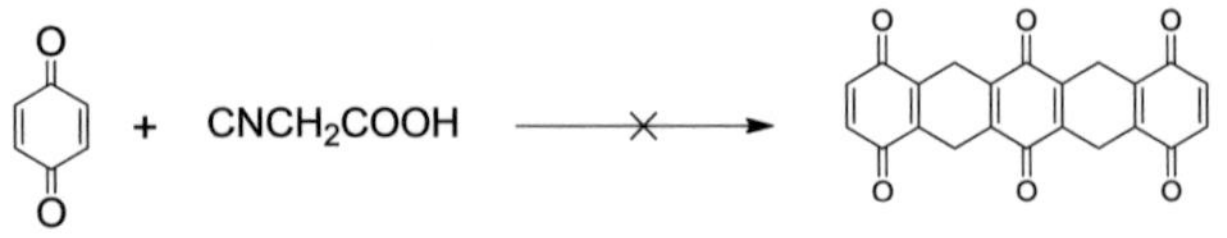

Scheme 2.5. Synthetic route to pentacenequinone according to Junek.

Greenhalgh et al. studied this product. They found that the product was not a 1,4-quinone and the structure of the pentacenequinone was untenable under this reaction condition.[62] The reaction of benzoquinone with cyanoacetic acid leads to a double cyclization, which resulted in a benzodihydrofuranone. Benzodihydrofuranone can be easily oxidized in the air, which leads to the conjugated molecule of benzodifuranone. The synthetic route and the suggested mechanism are described in Scheme 2.6.

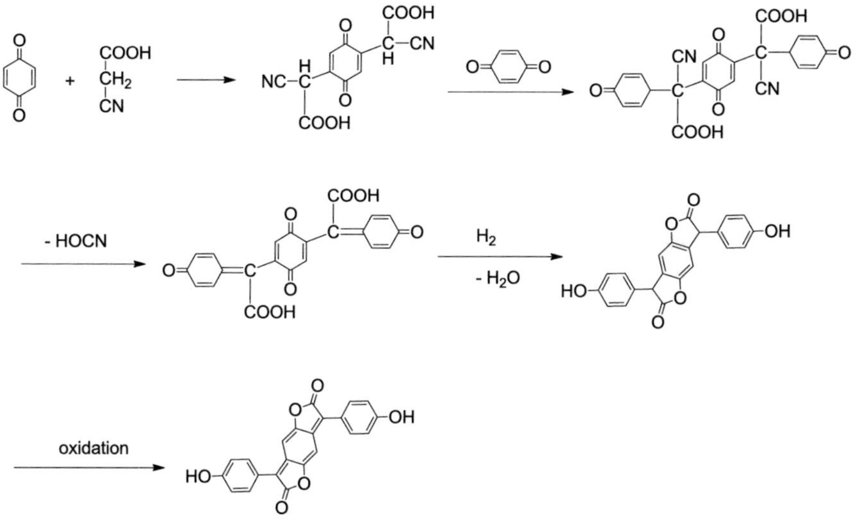

Scheme 2.6. Mechanism of the synthesis of benzodifuranone obtained from reaction of benzoquinone with cyanoacetic acid.

In the following studies, Greenhalgh et al. discovered that the reaction of benzoquinone or hydroquinone with mandelic acid derivatives results in the same class of benzodifuranone products (Scheme 2.7).[62, 63]

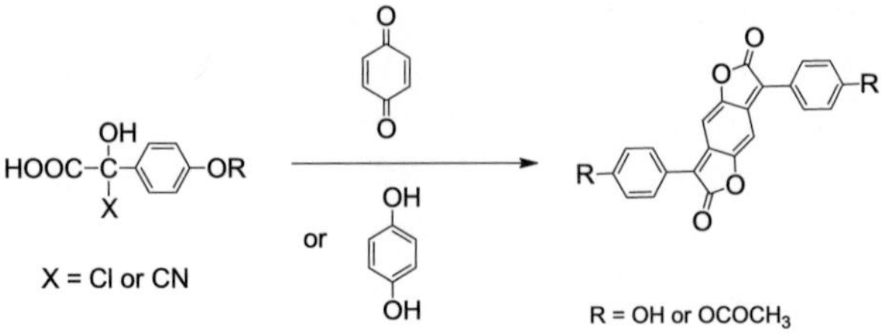

Scheme 2.7. Synthetic route to benzodifuranone according to Greenhalgh et al.

Since their discovery the benzodifuranone chromophore has been belonging to a class of currently used dyes and chromophores which have been developed in the last thirty years.[64] Because of their deep colour, good brightness of shape and light fashion, they were commercialized as disperse dyes for textiles, especially for polyesters. Depending on the substitution pattern, benzodifuranones exhibit red to blue colours,[63] and exhibit a strong solvatochromism in organic solvents.[65]

Among a handful published research works, Gorman et al. reported the solvatochromism of an aminobenzodifuranone (Fig. 2.12).[65] It was shown that with increasing polarity of the solvent, the aminobenzodifuranones exhibit positive solvatochromic shifts up to 200 nm.

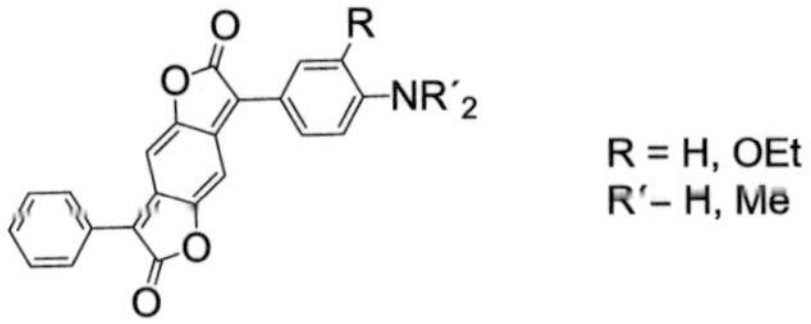

Fig. 2.12. Benzodifuranones containing amino groups.

Lately, Hallas and Yoon reported the synthesis and properties of a series of unsymmetrical benzodifuranones with colours in the range from yellowish red to bluish green.[66] The monomers were obtained in relatively good yields. They showed absorption maxima in the range from about 500 to 680 nm, exhibiting extinctions coefficients up to 50 000 Lmol⁻¹cm⁻¹. The structures of the unsymmetrical benzodifuranones are displayed in Fig. 2.13.

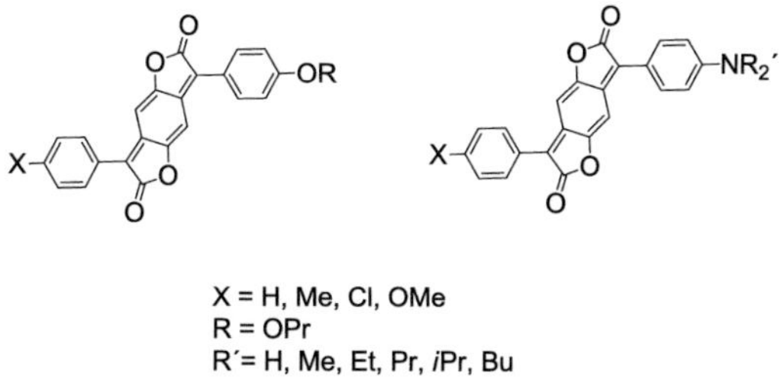

X = H, Me, Cl, OMe
R = OPr
R´= H, Me, Et, Pr, iPr, Bu

Fig. 2.13. Unsymmetrical benzodifuranones.

2.4 Metal-mediated and -catalyzed coupling reactions

For preparation of conjugated polymers metal-catalyzed cross-coupling reactions have been widely used. A vast number of this kind of reactions have been applied, such as nickel promoted coupling reactions (Yamamoto coupling[67, 68], Negishi coupling[69]), palladium catalyzed reactions (Heck coupling[70], Suzuki coupling[71], Stille coupling[72, 73], Sonogashira coupling[74-76], Buchwald-Hartwig coupling[40-45, 77]), and copper catalyzed reactions (Ullmann reaction[78], Glaser coupling[79]). Palladium-catalyzed polycondensations are regarded as the best methods. Especially Suzuki coupling, Stille coupling, Heck coupling and Sonogashira coupling are widely used in the field. The importance of Pd-catalyzed cross-coupling reactions was honered by awarding the Nobel Prize in chemistry in 2010 to R. Heck, E. Negishi and A. Suzuki. Some used cross-coupling reactions in this work are described in the following.

Suzuki coupling

Suzuki coupling is a Pd-catalysed coupling reaction for carbon–carbon (C–C) bond formation. The aromatic organoboron compounds serve as the active material to react with aromatic halides (Scheme 2.8).

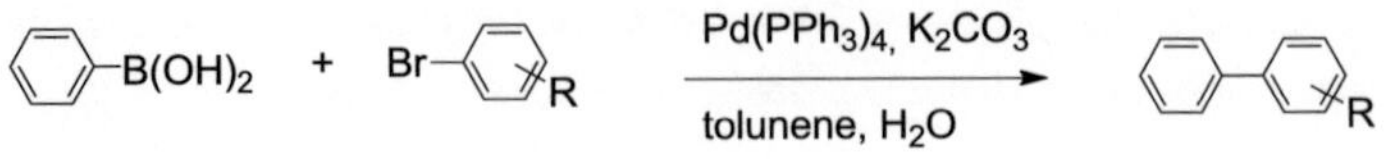

$\langle\text{Ph}\rangle\text{-B(OH)}_2$ + Br-$\langle\text{Ph}\rangle$-R →[Pd(PPh$_3$)$_4$, K$_2$CO$_3$][toluene, H$_2$O] $\langle\text{Ph}\rangle$-$\langle\text{Ph}\rangle$-R

Scheme 2.8. Suzuki coupling of phenylene derivatives.

Boronic acid may be replaced by potassium trifluoroborates, organoboranes or boronic esters. In place of the halide, the so-called pseudohalides such as triflates can also be used as coupling partners. One specialty of the Suzuki coupling is that the boronic acid or boronic acid ester must be activated, generally with a base, which is not necessary for the other cross-coupling reactions. The polarization of the organic ligand is increased after the activation of the boron atom, which leads to a transmetallation. A suggested mechanism of Suzuki coupling is shown in Figure 2.14.

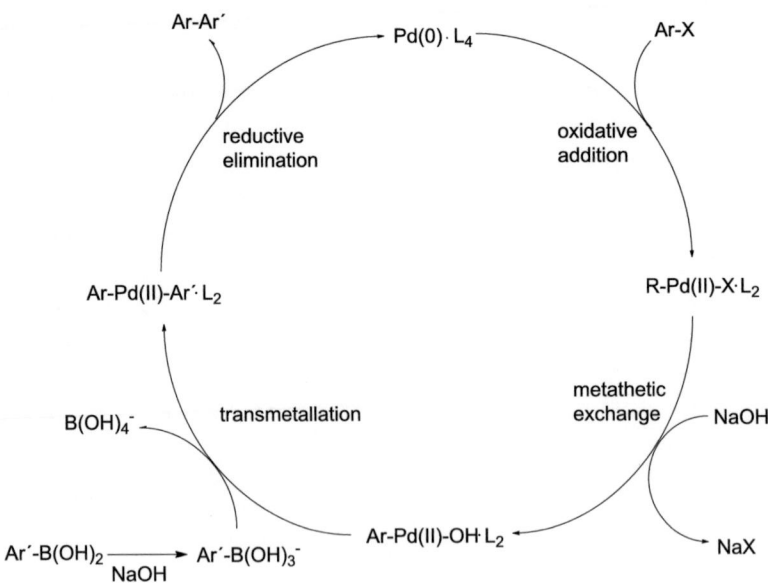

Fig. 2.14. Mechanism of Suzuki coupling.

At first, the oxidative addition of palladium(0) to the halide takes place, resulting in an organo-palladium(II) species. The intermediate compound Ar-Pd(II)-OH·L$_2$, which is formed by reaction of the organo-palladium(II) species with the base, transforms in a organo-palladium(II) species Ar-Pd(II)-Ar'·L$_2$ by transmetallation with the boronate complex. A following reductive elimination of Ar-Pd(II)-Ar'·L$_2$ gives the final coupling product and restores the original palladium catalyst from palladium(II). If the starting compounds, the boron acid and the halide, are bifunctional, a polymer can be obtained as product. In this work, preparation of bifunctional monomers is the first key research strategy.

Stille coupling

As a base-free method, Stille coupling offers an alternative to form C-C bonds (Scheme 2.9).

$$R\text{-}X + R'SnMe_3 \xrightarrow{Pd(PPh_3)_4} R\text{-}R' + XSnMe_3$$

Scheme 2.9. Stille coupling reaction.

A suggested mechanism of Stille coupling is described in Fig. 2.15. Similar to Suzuki coupling, the first step in the Stille coupling cycle is the oxidation of the palladium catalyst to the halide leading to a *cis* intermediate which rapidly isomerizes to the *trans* intermediate R$_1$-Pd(II)L$_m$-X. Transmetalation with the organostannane forms an intermediate R$_1$-Pd(II)L$_m$-R$_2$, which produces the final product R$_1$-R$_2$ and the active Pd(0) species after reductive elimination. The oxidative addition and reductive elimination retain the stereochemical configuration of the respective reactants.

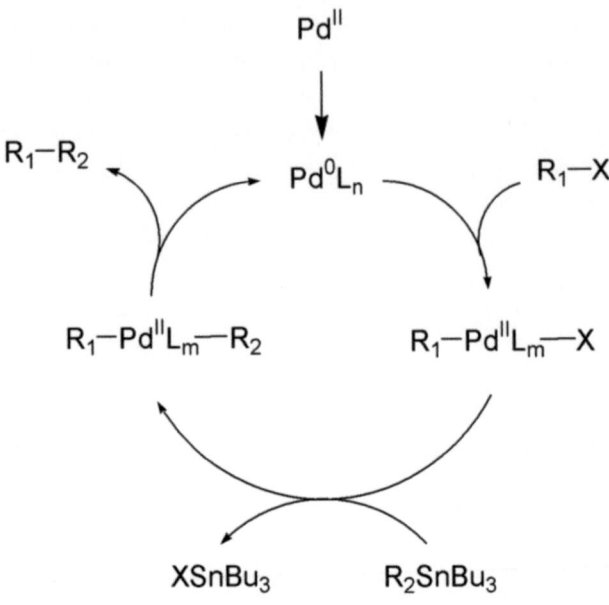

Fig. 1.25. Mechanism of Stille coupling reaction.

Sonogashira coupling

Sonogashira coupling is a coupling reaction of terminal alkynes with aryl or vinyl halides (Scheme 2.10).

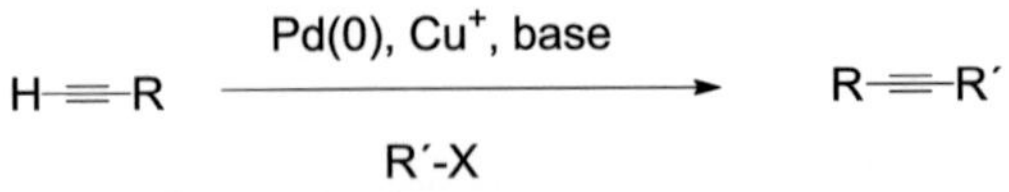

$$H\!\!=\!\!R \xrightarrow[R'\text{-}X]{Pd(0),\ Cu^+,\ base} R\!\!=\!\!R'$$

Scheme. 2.10. Sonogashira coupling reaction.

The reaction mechanism of Sonogashira is not clearly understood. It is suggested that the mechanism revolves around a palladium cycle (**A**) and a copper cycle (**B**).[80] The suggested mechanism is displayed in Fig. 2.16.

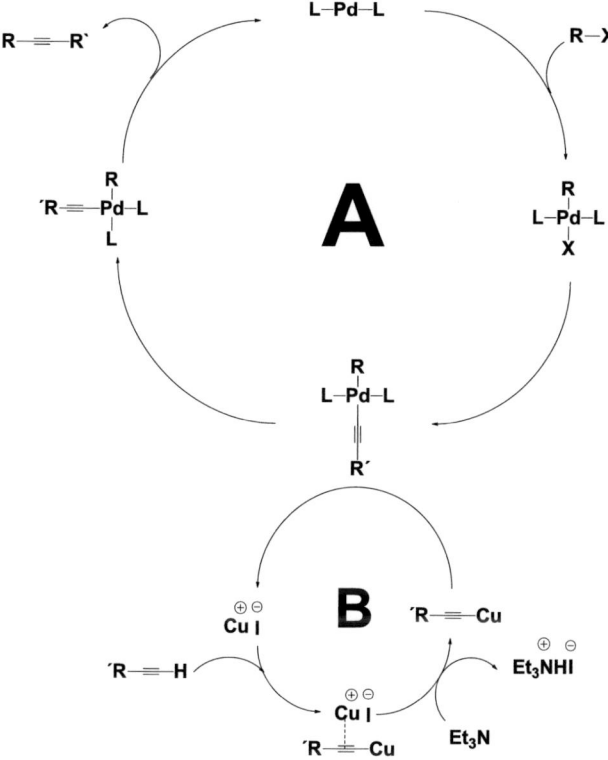

Fig. 2.16. Suggested mechanism of Sonogashira coupling.

First, in the palladium cycle (**A**), the active palladium catalyst Pd(0)·L$_2$ reacts with the aryl halide in an oxidative addition to yield the R-Pd(II)·X·L$_2$ complex, which reacts in a transmetallation reaction with the copper acetylide produced in the copper cycle (**B**). Expelling the copper halide CuX a complex R-Pd(II)-(C≡C)-R´·L$_2$ is formed. In the following a *trans-cis*-isomerization, the two *trans*-oriented ligands convert to a *cis*-formed complex. At the end the product R-(C≡C)-R´ is obtained in a reductive elimination with regeneration of Pd(0). At the same time, the reactivity of the triple bond is activated in the copper cycle (**B**). First, the deprotonation of the terminal alkyne takes place, although the employed amines such as triethylamine or *N,N*-diisopropylethylamine are not basic enough. However, the deprotonation is initiated after formation of a π-alkyne-copper complex. In the following, an organocopper compound R´-(C≡C)-Cu forms after reaction of the π-alkyne-cooper complex with the base and continues to react with palladium intermediate R-Pd(II)-(C≡C)-R´·L$_2$ under regeneration of copper halide CuX.

2.5 Electropolymerization

Since the first report on electropolymerized poly(pyrrole), electropolymerization has become a very important method to prepare conductive polymer films.[81] In the last twenty years, a vast number of works on electropolymerization of pyrrole derivatives have been published.[81-88] Despite the difficulty of preparation of pyrrole derivatives, a number of works on synthesis and characterization of pyrrole-based polymers using electropolymerization has been reported.[89-93] Fig. 2.17 outlines some pyrrole derivatives suitable for electropolymerization. In addition to the linear polymers, polymer network films containing pyrrole were also reported.[94]

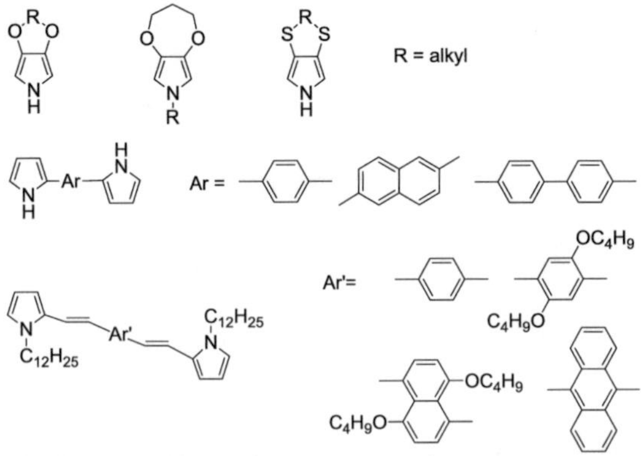

Fig. 2.17. Examples of pyrrole-based electropolymerizable monomers.

A much more popular electropolymerizable compound is thiophene, which has been widely reported in recent years. Thiophene can be easily connected with other arylenes, which is advantageous for formation of electropolymerizable monomers with different conjugation length. A so-called tuning of optical and electrochemical properties of the thiophene derivatives can be adjusted easily. The electropolymerization process of thiophene has been explored by many researchers.

A suggested mechanism of the electropolymerization is outlined in Fig. 2.18. First, the oxidation of a monomer produces a radical cation, which can then couple with a second radical cation to form a dication dimer, or with another monomer to produce a radical cation dimer. The dication dimer or radical dimer continue coupling with other thionphene units, resulting in polymeric chains. Deposition of long, well-ordered chains onto the electrode surface is followed by growth of either long, flexible chains, or shorter, more crosslinked chains, depending by the polymerization conditions. A number of techniques such as *in situ* video microscopy, cyclic spectrovoltammetry, photocurrent spectroscopy, and electrochemical quartz crystal microbalance measurements, have been used to elucidate the nucleation and growth mechanism leading to deposition of polymer onto the anode. Some examples of thiophene-containing monomers are shown in Fig. 2.19.

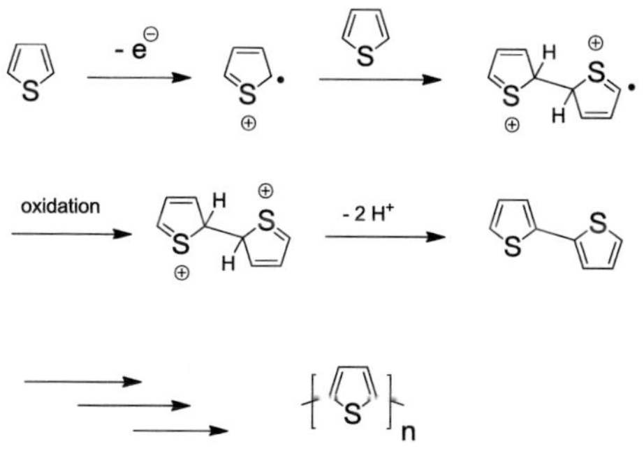

Fig. 2.18. Mechanism of electropolymerization of thiophene.

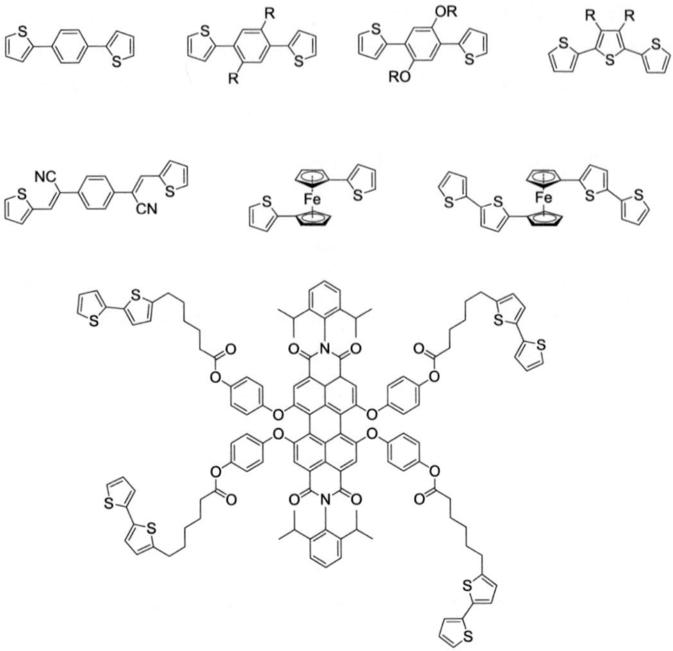

Fig. 2.19. Examples of electropolymerizable monomers containing thiophene.

Among the vast number of reports, Kabasakaloglu and colleagues studied the electrochemical properties of thiophene and polythiophene using different supporting electrolytes (ammonium perchlorate, tetraethylammonium tetrafluoroborate and tetrabutylammonium hexafluorophosphate) in acetonitrile.[95] The thiophene oxidation potential against Ag/Ag+ exhibited 1.6 V, using tetrabutylammonium hexafluorophosphate as the electrolyte. Pt has generally established itself as working electrode due to the fact that it adsorbs the weak acid anions such as PF_6^-, which benefits the polymer film formation process. A high-strength conducting polythiophene prepared in boron trifluoride diethyletherate solution was reported by Shi et al.[96] The polymer exhibited a tensile strength even greater than that of aluminium. Reynolds et al.[97] disclosed the synthesis of the bis-thiophene-arylenes and bis-furan-arylenes. The arylenes contain benzene, alkyl- and alkoxy-benzene. Lin et al. synthesized bis(2-cyano-2-α-thienylethenyl)arylenes later.[98] Destri and coworkers reported thiophene derivatives, which even become luminescent after electropolymerization.[99] As a stable chromophore, perylene was also coupled with thiophene derivatives and could be electropolymerized.[100]

Since the invention of 3,4-(ethylenedioxy)thiophene (EDOT),[101] the research work of electropolymerization has speeded up dramatically. EDOT is more stable than pyrrole, and is more easily oxidized than thiophene. The oxidation potential of EDOT is only 1.2 V (vs. Ag/Ag$^+$), which is much lower than that of thiophene (1.6 V vs. Ag/Ag$^+$). The lower oxidation potential originates from the electron-rich 3,4-ethylenedioxy group. The cation radical intermediates can be stabilized allowing the electropolymerization to proceed at low potentials and with a minimum of side reactions. The poly(3,4-(ethylenedioxy)thiophene) (PEDOT) film is easily formed at the anode. It exhibits an oxidation potential between -0.07 V and -0.28 V (vs. Ag/Ag$^+$) depending on the solvent system. The polymer films show an absorption maximum between 580 and 610 nm with a bandgap of 1.60 eV. After EDOT was first reported, similar compounds (3,4-(methylenedioxy)thiophene, 3,4-(propylenedioxy)-thiophene, 3,4-(butylenedioxy)thiophene were prepared by several researchers.[102, 103] Some examples are shown in Fig. 2.20.

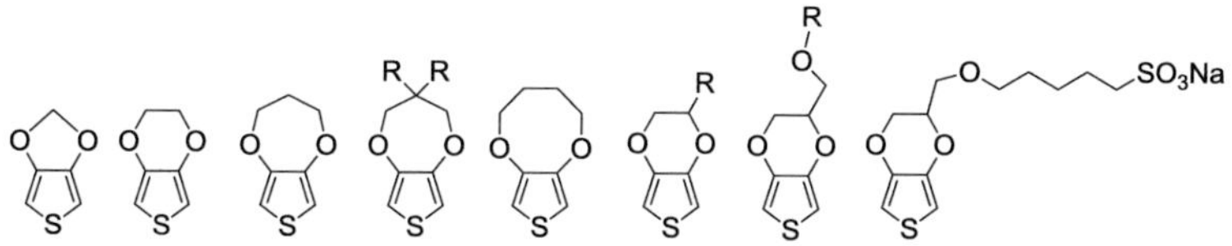

Fig. 2.20. EDOT and its derivatives.

As shown in Fig. 2.20, sulfonate substituted EDOT[104] and alkoxy derivatives[105] have been reported. Recently, sophisticated EDOT analogues have been synthesized. Roquet and Turbiez reported the synthesis of 3,4-phenylenedioxythiophene and the 3,4-ethylene-disulfanylthiophene derivatives.[106] All compounds can be easily electropolymerized and the resulting films exhibit electrochromic properties.

Similar to the sandwich-type bisthiophene arylene compounds, a large number of EDOT-arylene-EDOT monomers were reported. The same method as for thiophene derivatives to sandwich the arylene between two EDOT offers the opportunity to change the conjugation length of the monomer, and therefore to change the optical properties of the corresponding polymers. Compared with the bisthiophene arylene compounds, the EDOT-arylene-EDOT monomers can be oxidized at a lower potential and the possibility to incorporate various arylene units in the polymer backbone is increased. In Fig. 2.21 several EDOT-arylene-EDOT compounds are listed.

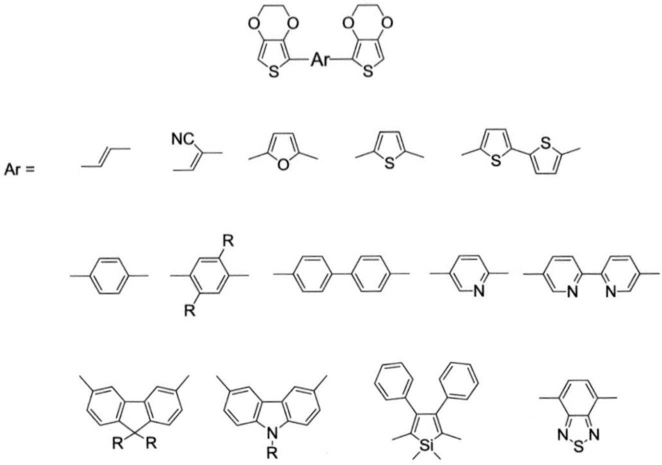

Fig. 2.21. Examples of sandwich-type EDOT-containing monomers.

Most of the bis-EDOT-arylene polymers exhibit ectrochromic properties. The colour change in response with the potential applied is useful in many applications. The water soluble PEDOT-PSS system was introduced by Bayer as Baytron P, which is deep blue in the neutral state and transparent in the oxidized state. The polymer was used as the active material for smart windows.[107] Reynolds et al. reported a dual polymer electrochromic device which was capable to switch between a colourless neutral state and a doped gray-green state.[108]

As the sulfur analogue of EDOT, 3,4-ethylenedithiathiophene (EDTT) (Fig. 2.23) was described by Kanatzidis et al in 1995.[109] An alternative preparation to various 3,4-bis(alkylthio)-thiophenes, including EDTT, was given by Meijer et al.[110] A combination of EDOT and EDTT, 3,4-ethylenedisulfanylthiophene (EDST) has also been reported (Fig. 2.23).[106, 109]

EDTT EDST

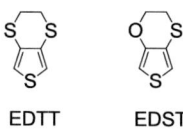

Fig. 2.23. Structures of EDTT and EDST.

Skabara et al. reported a series of well-defined EDTT-EDOT copolymers.[111] It was shown that the copolymers show strong intramolecular chalcogen–chalcogen interactions, which are responsible for persistent conformers in solution and solid state, resulting in significant influence on the properties of the materials. The results indicate that the S-O interaction is advantageous for the planarity of the polyEDOT chain.

There are some other units which can also be electropolymerized. Among those carbazole and triphenylamine are the most important ones (Fig. 2.24).

Carbazole Triphenylamine

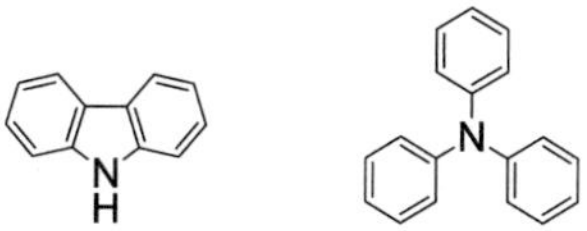

Fig. 2.24. Carbazole and triphenylamine.

Carbazole has been reported to be electropolymerizable at a very high potential (1.9 V vs. SCE).[112] Ma et al. reported on a luminescent polycarbazole film prepared by electropolymerization, which could be quite interesting for OLED devices.[113] Although triphenylamine itself cannot be electropolymerized, Chou et al. reported that starburst triarylamines can be electropolymerized.[114]

2.6 Electronic applications of conjugated polymers

2.6.1 Organic photovoltaic cells

There many application files for conjugated polymers. Organic photovoltaic cells (OPVC) belong to one of the most popular applications. They have attracted much attention in recent years. Among the extensive research activities, conjugated polymer-based solar cells have been widely studied.[9] The introduction of conjugated polymers for these application fields is advantageous to obtain cheap and easy methods to produce energy from light.[115] The general principle of organic photovoltaic cells (OPVC) is illustrated in Fig. 2.25.

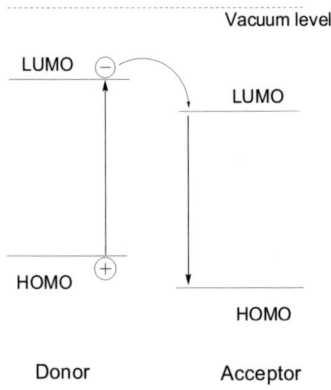

Fig. 2.25. Working principle of organic photovoltaic cells.

In general, bound electron-hole pairs (excitons) are produced after absorption of photons, which subsequently undergo dissociation. The electrons are excited to the lowest unoccupied molecular orbital (LUMO) and leave holes in the highest occupied molecular orbital (HOMO) forming excitons. The potential created by the different work functions helps to separate the exciton pairs, pulling electrons to the positive electrode and holes to the negative electrode. The current and voltage result in electricity.

There are different types of organic photovoltaic cells. Among various types of cells the simplest one is the single layer organic photovoltaic cell. A basic structure of this cell is illustrated in Fig. 2.26a. In the cell a layer of organic electronic materials is sandwiched between two metallic conductors, typically a layer of indium tin oxide (ITO) with high work function, and a layer of a low work function metal such as Al, Mg and Ca.

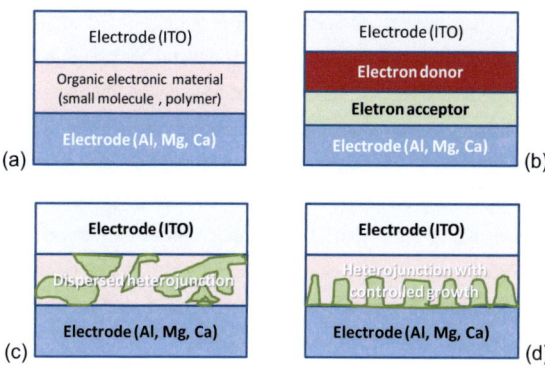

Fig. 2.26. Different types of organic photovoltaic cells (OPVC) with single layer (a), bilayer (b), dispersed heterojunction layer (c), and heterojunction layer with controlled growth (d).

In practice, single layer organic photovoltaic cells only show low quantum efficiencies (<1%) and low power conversion efficiencies (<0.1%).[116, 117] A major problem is the electric field resulting from the difference between the two conductive electrodes, which seldom is sufficient to break up the photogenerated excitons. The electrons recombine with the holes rather than reach the electrode. To solve this problem, multilayer organic photovoltaic cells were developed (Fig. 2.26b). A bilayer heterojunction configuration containing a p type layer for hole transport and an n-type layer for electron transport has been implemented by Tang to improve the photocurrent of the solar cell device.[118] The bilayer organic photovoltaic cells contain two different layers between the electrodes. The materials of the layers exhibit different electron affinity and ionization energy. Therefore electrostatic forces are generated at the interface between the two layers. The excitons are separated much more efficiently than in the single layer photovoltaic cells. Usually the layer with higher electron affinity and ionization potential functions as the electron acceptor. The other layer with lower electron affinity and ionization potential functions as the electron donor. This structure type is also called planar donor-acceptor heterojunction.[119] Due to the fact that the limited lifetime only allows excitons to diffuse a short distance, between 5 and 14 nm,[120-124] donor excitons created too far away from the heterojunction interface decay to the ground state without reaching the acceptor. This leads to the loss of absorbed photons and quantum efficiency. Consequently, the performance of bilayer heterojunction devices is greatly limited by the small area of the charge-generating interface between the donor and acceptor. A solution to overcome this difficulty is the concept of a bulk heterojunction (BHJ). Bulk heterojunctions (BHJ) were first introduced by the pioneering work of Yu et al.[4] Donor (D) and acceptor (A) materials were blended together, an interpenetrating network with a large D-A interfacial area was achieved through controlling the phase separation between the two components in bulk. This principle allows that the absorbing site in the composite is within a few nanometers of the donor–acceptor interface, leading to much enhanced quantum efficiency of charge separation. The formation of a bicontinuous network creates two channels to transport holes in the donor domain and electrons in the acceptor domain, resulting in efficient charge collection (Fig. 2.26c). Sariciftci and co-workers achieved a major breakthrough and rapid development of BHJ solar cells arose from the discovery of efficient photoinduced electron transfer in conjugated polymer–fullerene composites.[115] Buckminsterfullerene (C_{60}) has proven to be an ideal n-type material due to its various intrinsic advantages (Fig. 2.27). However, the tendency to crystallize and the poor solubility of C_{60} in organic solvents hinder direct applications in inexpensive solution-based processing techniques.[125] In 1995, Wudl and Hummelen synthesized one particular solubilizing derivative of C_{60}, [6,6]-phenyl-C_{61}-butyric acid methyl ester (PCBM), which is the most ubiquitously used acceptor for current BHJ solar cell research.[125] Recently, a C_{70} derivative, C_{70} PCBM ($PC_{71}BM$), has been synthesized (Fig. 2.27).[126, 127] It has been shown that the efficiency of BHJ devices can be further improved by replacing acceptor C_{60} PCBM with its higher fullerene analogue which has lower symmetry and allows more transitions.

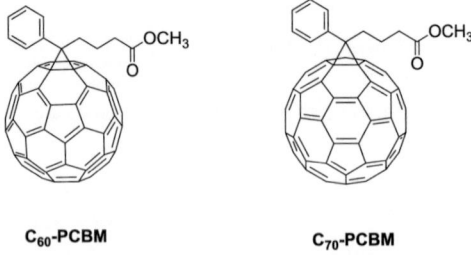

C_{60}-PCBM C_{70}-PCBM

Fig. 2.27. Molecular structures of C_{60}-PCBM and C_{70}-PCBM.

A number of reviews have examined the use of conjugated polymers in solar cell applications.[128-132] Over the past decade a tremendous research effort has created numerous novel conjugated polymers. However, the device performance in BHJ solar cells has been steadily enhanced and power conversion efficiencies (PCE) higher than 2% are becoming more and more commonplace. In the last three years, a PCE of over 5% has been achieved for many newly developed low band gap conjugated polymers. Recent research work of Bazan et al. reported a polymer based on dithienosilole and benzothiadiazole, achieving a PCE of 5.9%.[133] A very recent publication by Zhao et al. reported polymer solar cells with a PEC of 6.5%.[134] The polymer is based on poly(3-hexylthiophene) (P3HT) as donor and an indene-C_{60} bis-adduct (ICBA) as acceptor with an open-circuit voltage of 0.84 V, a short-circuit current of 10.61 mA/cm^2, and a fill factor of 72.7% under irradiation at 100 mW/cm^2.

2.6.2 Electroluminescent applications

Photoluminescence of π-conjugated molecules was found in the middle of the 19th century. However, electroluminescence stayed unknown for almost 100 years. The first observation of electroluminescence in organic materials was in the early 1950s by Bernanose and co-workers. They applied high-voltage alternating current (AC) fields in air to materials such as acridine orange, either deposited on or dissolved in cellulose or cellophane thin films.[135-137] The first diode device was reported Tang and van Slyke in 1987.[138] This resulted in a reduction in operating voltage and improvements in efficiency and led to the current era of OLED research and device production.

Polymer light-emitting diodes (PLED) contain an electroluminescent conductive polymer, which emits light when connected to an external power source. They can be used as a thin film for full-spectrum colour displays. PLEDs are quite efficient and require a relatively small amount of power for the amount of light produced. In 1990, Burroughes et al. reported a high efficiency green light-emitting polymer-based device using 100 nm thick films of poly(p-phenylene vinylene).[139]

The structure of an electroluminescent device based on conjugated polymer is illustrated in Fig. 2.28. The principles of electroluminescent devices are described as follows:

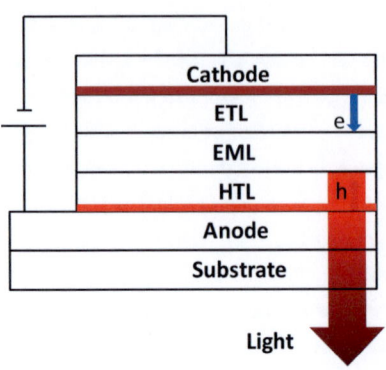

Fig. 2.28. A typical electroluminescent device.

Between the metallic cathode, usually Al or Ca, and the anode on a tranparent substrate, usually ITO, several layers of electron transport layer (ETL), emitting materials layer (EML), and hole transport layer (HTL) are sandwiched. Under an applied voltage, negative charges such as electrons from the cathode and positive charges such as holes from the anode are injected into the layers. Exitons are formed through electrostatic interaction between the electrons and holes, which, at the end, generates light through the radiative recombination of excitons. Some examples of polyfluorene-based emitters are listed in Fig. 2.29.

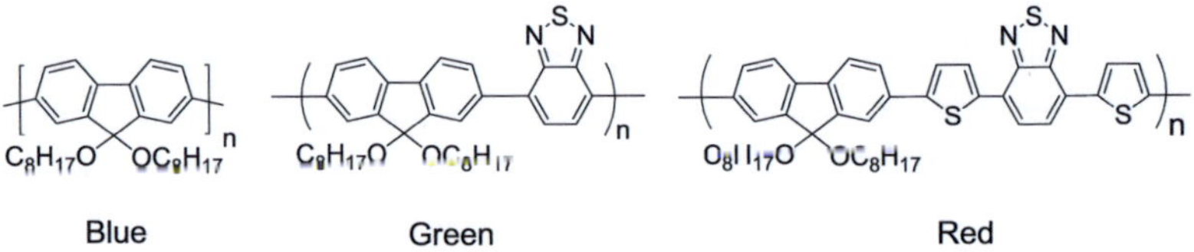

Blue   Green   Red

Fig. 2.29. Polyfluorene-based emitters with different colours.

2.6.3 Electrochromism

Electrochromic materials have gained considerable interest due to their applicability in many devices as smart windows, optical shutters, displays, and mirror devices. Electrochromism is defined as a reversible change of the optical absorption of a material induced by an external voltage, with many inorganic and organic species showing electrochromism throughout the electromagnetic spectrum. In the recent years, conjugated polymers have become a new class of electrochromic (EC) materials due to their ease of processability, rapid response times, high optical contrast, and the ability to modify their structure to create multicolour electrochromism. A typical conjugated polymer based EC device is shown in Figure 2.30.

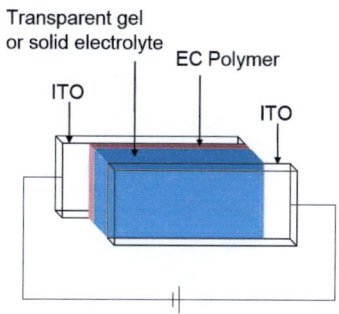

Fig. 2.30. A typical electrochromic device.

By oxidation (p-doping) and reduction (n-doping), the band structure of the neutral polymer is modified, lower energy intraband transitions and charge carriers such as polarons and bipolarons are generated, which are responsible for increased conductivity and change of optical absorption. These kind of optical and structural changes are reversible through repeated doping or dedoping over many redox cycles. This makes the conjugated polymers potentially useful in many applications. Lately, devices build on conjugated polymers such as poly(thiophene), poly(pyrrole), and poly(aniline) have been widely studied.[140] Among those research activities, Reynolds et al. reported a black-to-transmissive and multicoloured switching electrochromic device (ECD) constructed in a pseudo 3-electrode electrochromic device (P3-ECD) architecture.[141]

2.7 Conjugated microporous polymers

Microporous organic polymers (MOP), such as polymers with intrinsic microporosity (PIMs),[142-145] hyper-cross-linked polymers (HCPs),[146-148] covalent organic frameworks (COFs),[149, 150] have become of great interest in diverse application areas, such as gas separation,[151] gas storage,[147, 148, 151, 152] or heterogeneous catalysis.[151, 153, 154] The application performance of these materials depends on the micropore structure and the physical surface area. The highest apparent Brunauer–Emmett–Teller (BET)[155] surface areas achieved in amorphous MOPs are of the order of 2000 $m^2 g^{-1}$.[156] Recently, organic microporous materials have been studied intensely. According to their controllable synthesic route, this class of materials has been gaining a lot of interest. In particular, the introduction of conjugated units into microporous networks has been gaining more and more attention lately.[157-159] Beside the usual application potentials for high surface area materials, for example gas storage or catalysis matrix support, conjugated polymer networks might be used in advanced fields such as electrical conductivity, light emission, and UV/vis absorption.

In recent years, a number of research works on conjugated microporous polymers have been published due to the fact that they provide the potential to combine high surface areas with physical properties relevant to organic electronics. The first examples of microporous polymers (CMPs) with a fully conjugated system exhibiting high surface areas in the dry state was reported by Jiang et al. in 2007. The polymers consist of poly(aryleneethynylene) (PAE) with BET surface areas up to 834 m^2g^{-1}. In comparison, a hyper-crosslinked carbinol-containing polymer showed a BET surface area up to 1000 m^2g^{-1}. However, the conjugation is broken by the carbinol centre in the repeat unit. Kijima et al. reported preparation of microporous pyrolytic polymers by heating branched alkyne-linked networks to temperatures in the range from 350 to 900 °C.[160] Using Sonogashira-Hagihara cross-coupling polycondensation reactions, a series of PAE-containing CMPs were prepared by Jiang et al. (Fig. 2.31).[159, 161] It was shown that the micropore size, the micropore volume, and the BET surface area of the networks could be tuned by varying the "strut"-length of the monomers. Jiang et al. also reported preparation of CMPs using Cu-catalyzed homocoupling of phenylene butadiynylene.[162] The networks exhibited high surface areas ($S_{BET} > 800$ m^2g^{-1}). Microporous polysilanes were reported recently.[163] The materials exhibits BET surface areas up to 1046 m^2g^{-1}. Fig. 2.31 shows a series of CMP networks with various "strut" lengths.

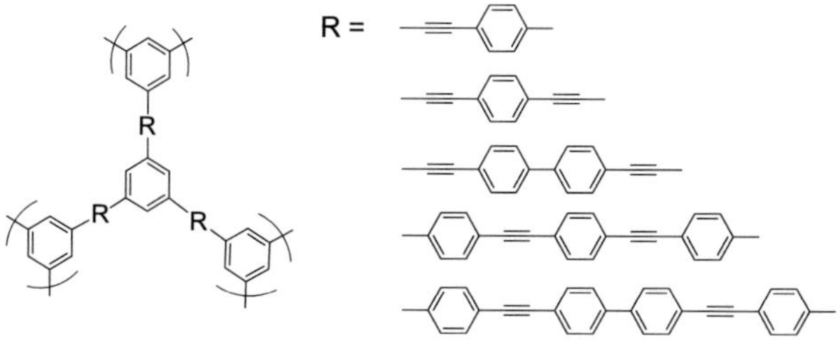

Fig. 2.31. CMP networks prepared using Sonogashira-Hagihara cross-coupling with various "strut" lengths. The BET surface area varies from 512 to 1046 m^2g^{-1}.[159]

The synthesis and application of conjugated microporous polymers is a novel research area, which may also open doors to electronic application fields. However, the challenge is still the insolubility of the materials, which complicates the fabrication process.

BET isotherm

To understand the surface area properties of conjugated microporous polymers, a brief overview of some analytic methods is necessary. The common method to analyze the surface area is the gas sorption measurement, in which gas molecules are adsorbed and desorbed. This process is called adsorption. In 1916, Langmuir[164] presented a semi-empirical model for gases adsorbed to solids. He assumed that every adsorption site is equivalent, and that the ability of a particle to bind depends on whether the nearby sites are occupied. The dynamic equilibrium of the free and adsorbed gas is described as:

$$A_{gas} + S \rightleftharpoons AS \qquad (1)$$

where A is the adsorption and S is the surface. The rate of increase of surface coverage is proportional to the pressure. This model, the so-called Langmuir isotherm is suitable for gas adsorption taking place in monolayers. However, in the practice, adsorption usually proceeds in multilayers and the first adsorbed layer may act as a substrate for further adsorption. It cannot be ignored. The most widely used isotherm dealing with multilayer adsorption was derived by S. Brunauer, P. Emmett and E. Teller in 1938, and it is called the BET isotherm.[155] It deals with the assumption that the adsorption takes place in a homogenous layer, and the gas molecules are adsorbed without interact with each other with a constant adsorption enthalpy $\Delta_{ad}H$. Every first adsorbed layer acts as a substrate for the second adsorbed layer. The adsorption enthalpy of the second and further layers equals the condensation enthalpy $\Delta_{cond}H$.

The BET isotherm is described as:[165]

$$\frac{V}{V_m} = \frac{CP}{[P_0 - P] + (C-1)P/P_0} \quad \text{with} \quad C = e^{-(\Delta_{ad}H^\ominus + \Delta_V H^\Delta)/RT} \quad (2)$$

and V being the volume of adsorbed vapour at pressure P, V_m the volume of gas adsorbed when the entire surface is covered by a monomolecular layer, C the BET constant, P the pressure, and P_0 the saturation vapour pressure. The shape of the BET isotherms is illustrated in Fig. 2.32.

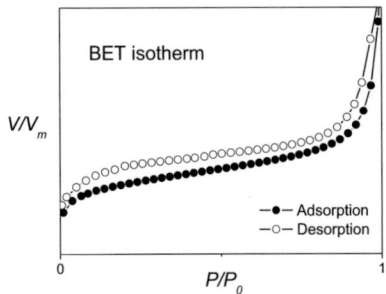

Fig. 2.32. Plots of the BET isotherms.

Equation (2) can be written in linear form as:

$$\frac{P}{V(P_0 - P)} = \frac{1}{V_m C} + \frac{(C-1)P}{V_m C P_0} \quad (3)$$

As shown in Fig. 2.33, a plot derived from the expression $P/[P_0 V(1-P/P_0)]$ vs. P/P_0 is called the BET plot, and a straight line can be obtained in the region of relative pressure P/P_0 near complete monolayers (0.05 ≤ P/P_0 ≤ 0.35). From the slope $(C-1)/V_m C$ and the y-intercept $1/V_m C$, the volume V_m of the gas adsorbed can be calculated. The total surface area of the adsorbents can be calculated from V_m, if we know the space requirement of the single adsorbed molecule. For example, the space requirement of the nitrogen molecule is about

$16.2 \cdot 10^{-20}$ m². So a specific surface area of 560 m²g⁻¹ on a substrate of silica gel can be calculated.

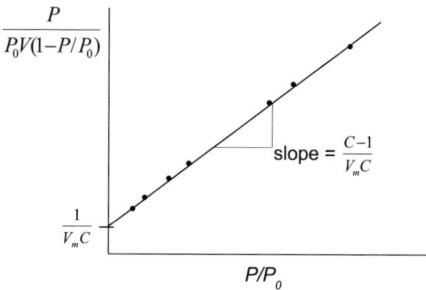

Fig. 2.33. Illustrated BET plot, derived from equation (3).

Barrett-Joyner-Halenda (BJH) method

In 1951, E. Barrett, L. Joyner and P. Halenda described a technique to determine the pore volume and area distribution of porous adsorbents to molecules of various size.[166] This technique is called BJH method. It is assumed that pores are cylindrical in shape, and the radius of the capillary is equal to the sum of the Kelvin radius and the thickness of the adsorbed layer on the pore walls. It is also assumed that the thickness of the adsorbed layer on a flat surface is the same as the thickness t of the adsorbed layer inside the pore, and all pores are filled at a relative pressure of $P/P_0 = 1.0$. In fact, the BJH technique divides the pore sizes into groups, and for simplicity it is assumed that all pores in each group of capillaries have an average radius. In general, it is recommended that the average pore radius increases by 0.5 nm from one pore size group to the other up to a pore radius of about 6 nm. For pore radii over 6 nm, it increases by 1 nm. The desorption branch of the isotherm obtained experimentally is used for the analysis. Computations start at a relative pressure of 0.967 down to a relative pressure of ca. 0.4. The BJH method considers the isotherm as a series of steps, when the pressure is lowed. The amount of adsorbent removed during each step is given by the ratio of the processes of emptying inner pore and film thinning. For each group of pores, the pore core is calculated, as well as the change in thickness of the adsorbed film during each step of pressure reduction. The total pore volume and the surface areas of pores can be easily calculated by adding the values calculated in each pore groups.

t-plot method

From 1964 to 1965, Lippens and de Boer developed a method for analysis of pore structure in a series of publications.[167, 168] This method is called the t-plot method, which has attracted great attention as a simple and direct interpretation of nitrogen isotherms. The values of relative pressure, P/P_0, measured during the experiment can be transformed to values of thickness t of the adsorbed layer. The plot of the volume V adsorbed vs. the thickness t of the adsorbed layer can be obtained. This plot is called a t-plot. For multimolecular adsorption, the experimental points should fall in a straight line and pass through the origin for a non-porous material. From the slope (V/t) of this line the surface area S can be empirically described as: $S = 15.47$ (V/t).[167-169] For a porous material, the line will have a positive intercept indicating micropores, or deviate from linearity suggesting filling of mesopores. t-Plots of a porous and non-porous material are illustrated in Fig. 2.34.

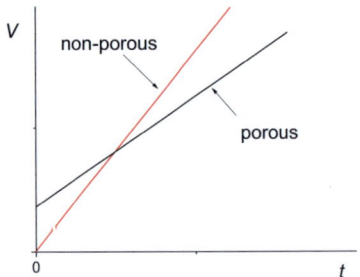

Fig. 2.34. Illustrated t-plots.

In 1944, the thickness t of the adsorbed monolayer was first proposed by Harkins and Jura to depend on P/P_0 accoring to the empirical equation as:[170]

$$t = \sqrt{\frac{13.99}{0.034 - \lg(P/P_0)}} \qquad (4)$$

Later in 1948, G. Halsey[171] proposed an empirical equation for the thickness of an adsorbed film on a pore wall, which has generally been accepted. The equation is described as:

$$t = 3.54 \sqrt[3]{\frac{-5}{2.303 \lg(P/P_0)}} \tag{5}$$

where the thickness t is given in Ångström (Å).

3 Conjugated polymers based on diketopyrrolopyrrole (DPP)

In this chapter, synthesis and characterization of new conjugated polymers based on diketopyrrolopyrrole (DPP) are described.

3.1 2,5- and 3,6-Di-substituted pyrrolo[3,4-c]pyrrole-1,4(2H,5H)-dione (DPP)-based conjugated linear polymers

Electropolymerization represents an alternative route to prepare conjugated polymers in a catalyst-free approach.[172-175] In recent publications of our group, electropolymerization of a bis-EDOT-substituted N-alkylated DPP monomer was studied.[176] An electrochromic polymer with low oxidation potential and reversible switching behaviour was obtained.

The purpose of the present chapter is to study the electropolymerization of bis-EDOT-substituted DPP-derivatives in greater detail. Special emphasis is paid to different substitution patterns of the DPP core chromophore, and the influence of the substitution pattern on the electropolymerization, and the optical and electrochemical properties of the resulting polymer films. It was interesting to study potential differences in electropolymerization and properties of deposited polymer films, if DPP monomers with EDOT units either attached to the phenyl groups in 2- and 5-position, or 3- and 6-position of the central chromophore are used (Fig. 3.1).

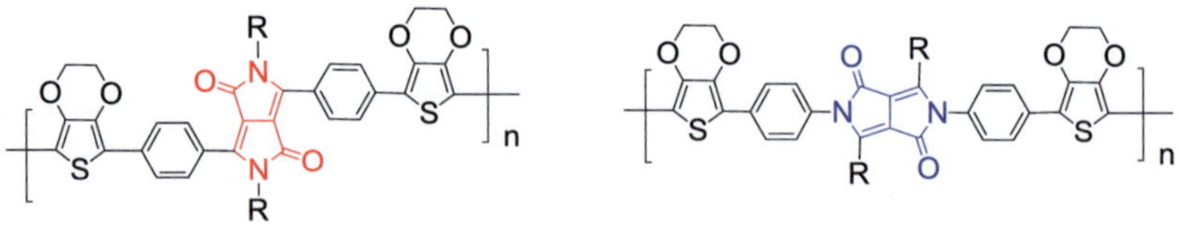

Fig. 3.1: Electropolymerized EDOT-DPP-polymers, R = aryl or alkyl.

Furthermore, we either wanted to study the role of an N-alkylation or N-arylation of the DPP unit on the properties of the resulting polymer films. The polymers exhibit interesting electrochemical and optical properties, which have been studied in detail using cyclic voltammetry and spectroelectrochemical analysis.

3.1.1 Synthesis and properties of monomers

The synthetic routes to the three monomers are displayed in Scheme 3.1. Different monomers were prepared containing EDOT units either attached to phenyl rings in the 3- and 6-position of DPP (**EDOT-DPP1a** and **EDOT-DPP1b**), or to phenyl rings in the 2- and 5-position (**EDOT-DPP2**). **EDOT-DPP1a** and **1b** differed in the substitution of the lactam group either by 2-hexyldecyl groups (**EDOT-DPP1a**) or 4-t-butylphenyl units (**EDOT-DPP1b**). The starting compounds **BrDPP**, **BrDFF** and **tBuDFF** were prepared according to previously reported procedures.[13, 21, 35] **BrDPP1a** was prepared using alkylation of 3,6-bis(4-bromophenyl)-pyrrolo[3,4-c]pyrrole-1,4(2H,5H)-dione (**BrDPP**) with 1-iodo-2-hexyldecane in 35% yield. **BrDPP1b** was obtained by condensation of 3,6-bis(4-bromophenyl)furo[3,4-c]furan-1,4-dione (**BrDFF**) with 4-tert-butylaniline in 60% yield, and BrDPP2 was prepared by condensation of 3,6-bis(4-(tert-butyl)phenyl)furo[3,4-c]furan-1,4-dione (**tBuDFF**) with 4-bromoaniline in 55% yield, respectively. The electropolymerizable monomers **EDOT-DPP1a**, **1b** and **2** were synthesized by Stille coupling (Scheme 3.1) of **BrDPP1a**, **BrDPP1b** and **BrDPP2** with (2,3-dihydrothieno[3,4-b][1,4]dioxin-5-yl)trimethylstannane (**EDOT1**) using Pd(PPh$_3$)$_4$ as catalyst and DMF as solvent under microwave conditions. The coupling reaction under microwave conditions gave high yields of 90, 85 and 82% for **EDOT-DPP1a**, **1b** and **2**, respectively. **EDOT-DPP1a**, **1b** and **2** are dark red solids, which are soluble in common organic solvents such as chloroform, DCM, DMF, THF and toluene, for example.

The ^{1}H-NMR spectra of **EDOT-DPP1a**, **1b** and **2** display all the expected resonances with no discernible peaks corresponding to impurities (see Chapter 5, Experimental Part). Significantly, the singlet signal at about 6.4 ppm can be ascribed to the H-atom of the thiophene unit. The two triplet signals with chemical shift at about 4.34 ppm are typical for the ethylene bridge of the EDOT unit. The other signals are very similar to those of the starting compounds **BrDPP1a**, **BrDPP1b** and **BrDPP2**.

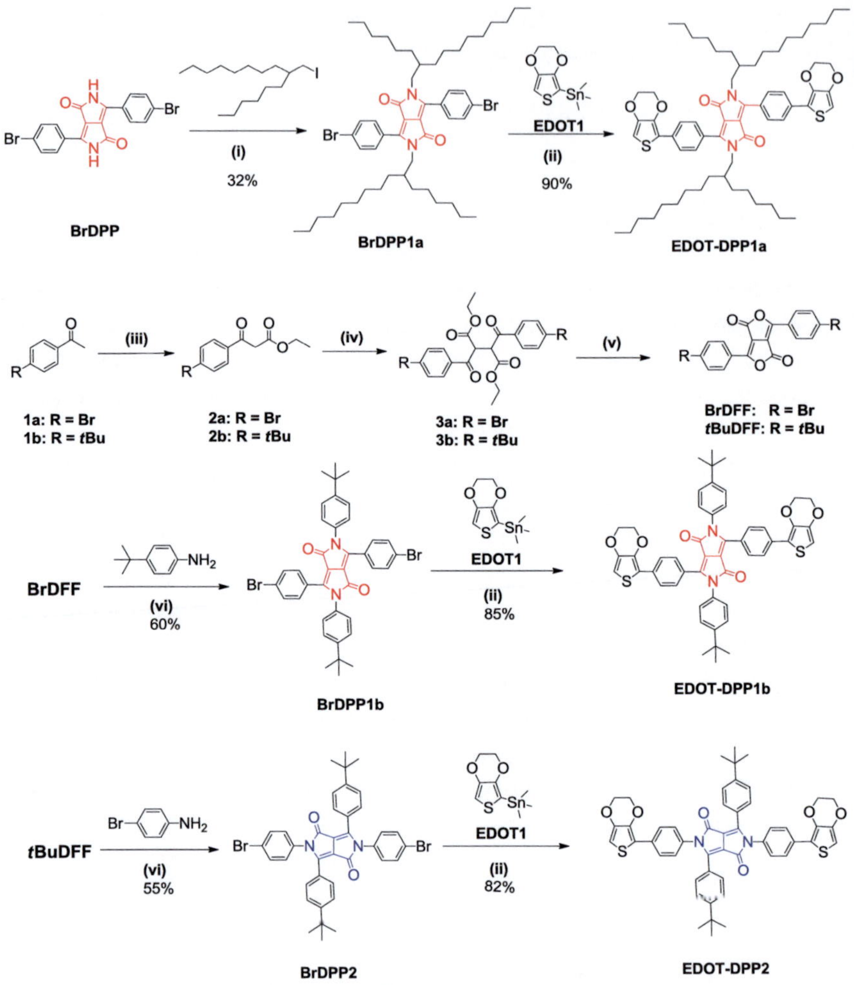

Scheme 3.1: Reagents and conditions: (i) KOH, DMSO, 190 °C, over night; (ii) Pd(PPh$_3$)$_4$, DMF, microwave, 160 °C, 1 h; (iii) diethyl carbonate, NaH, toluene, reflux; (iv) (a) Br$_2$, CHCl$_3$; (b) NaH, diethyl ether, reflux; (v) 200 °C, 20 min; (vi) N,N'-dicyclohexylcarbodiimide, CF$_3$COOH, CHCl$_3$, 3 d.

In the UV/vis absorption spectra (Fig. 3.2) of **EDOT-DPP1a** and **1b** indicate that a bathochromic shift of 20 to 30 nm is introduced by the addition of EDOT into the conjugated system of the starting compounds **BrDPP1a** and **BrDPP1b**. In DCM, **EDOT-DPP1a** exhibits absorption and emission maxima at 500 and 581 nm, and **EDOT-DPP1b** at 532 and 585 nm, respectively. **EDOT-DPP 2** with absorption and emission maxima at 492 and 526 nm shows no bathochromic shift compared with the corresponding compound **BrDPP2**. The reason is that the addition of EDOT to the N-phenyl group of **BrDPP2** does not affect the π-conjugation of the DPP unit because the N-phenyl group is separated from the conjugated system by the lactam N-atom.

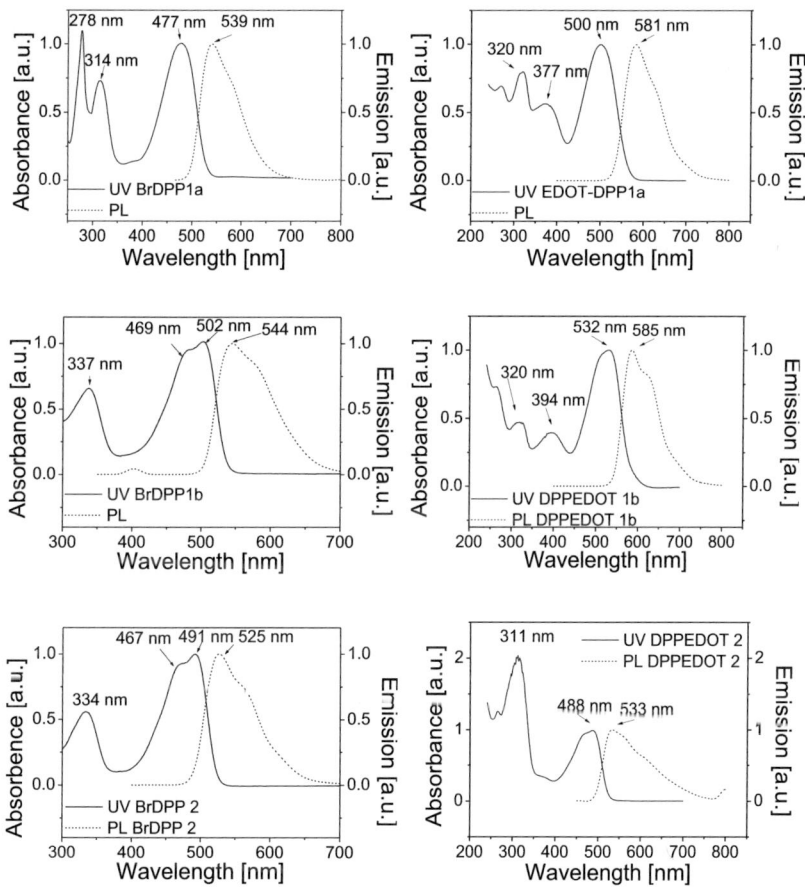

Fig. 3.2: UV/vis absorption and fluorescence spectra of **BrDPP1a**, **BrDPP1b** and **2**, **EDOT-DPP1a**, **1b** and **2**.

Absorption and emission colours of the brominated and the EDOT-substituted DPPs in dichloromethane are shown in Fig. 3.3.

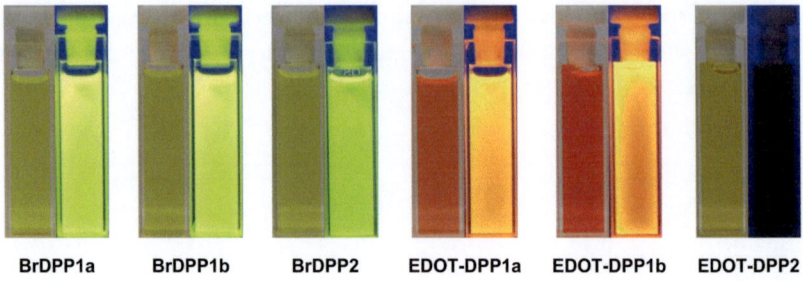

| BrDPP1a | BrDPP1b | BrDPP2 | EDOT-DPP1a | EDOT-DPP1b | EDOT-DPP2 |

Fig. 3.3. Absorption and emission colours of **BrDPP1a, BrDPP1b and BrDPP2, EDOT-DPP1a, 1b** and **2** in DCM solution.

3.1.2 Electropolymerization and properties of polymers

The polymers were prepared via anodic oxidative polymerization into thin films on ITO-coated glass substrates (Scheme 3.2).

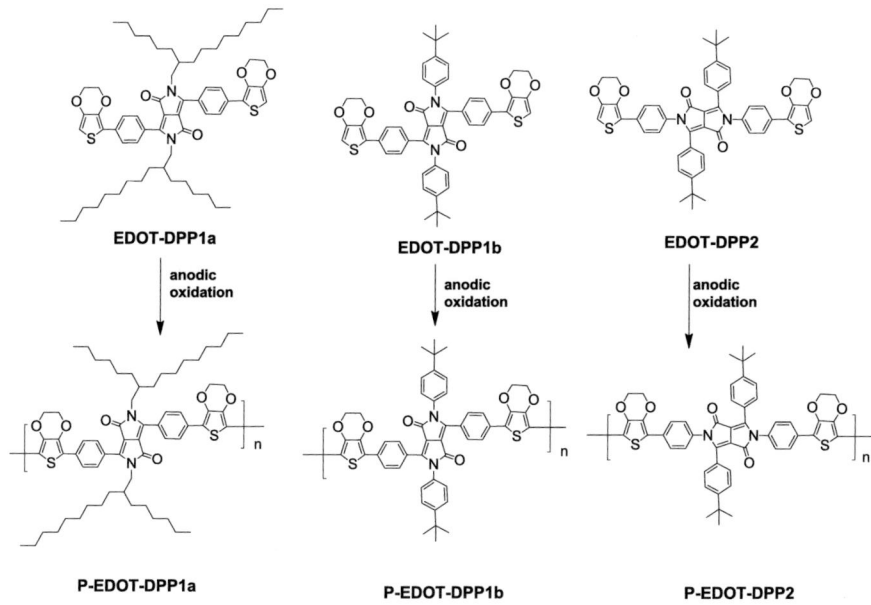

Scheme 3.2: Electropolymerization of **EDOT-DPP1a, 1b** and **2** by anodic conditions.

Electrochemistry of the monomers

In Fig. 3.4, potential cycles of the monomers in 0.1 M TBAPF$_6$ in DCM are shown. The oxidation potentials of the monomers were 0.64 V for **EDOT-DPP1a**, 0.72 V for **EDOT-DPP1b** and 0.83 V for **EDOT-DPP2**. This was expected because our previous study of the oxidation potential of an EDOT-substituted DPP monomer gave a similar value (0.70 V).[170] All the three monomers show quasi-reversible cathodic waves. For **EDOT-DPP1a**, the reductive cycle shows a cathodic wave at -1.69 V, which is reverted at -1.60 V. **EDOT-DPP1b** shows a cathodic wave at -1.58 V, which is reverted at -1.46 V. **EDOT-DPP2** shows a cathodic wave at -1.83 V, which is reverted at -1.66 V, respectively.

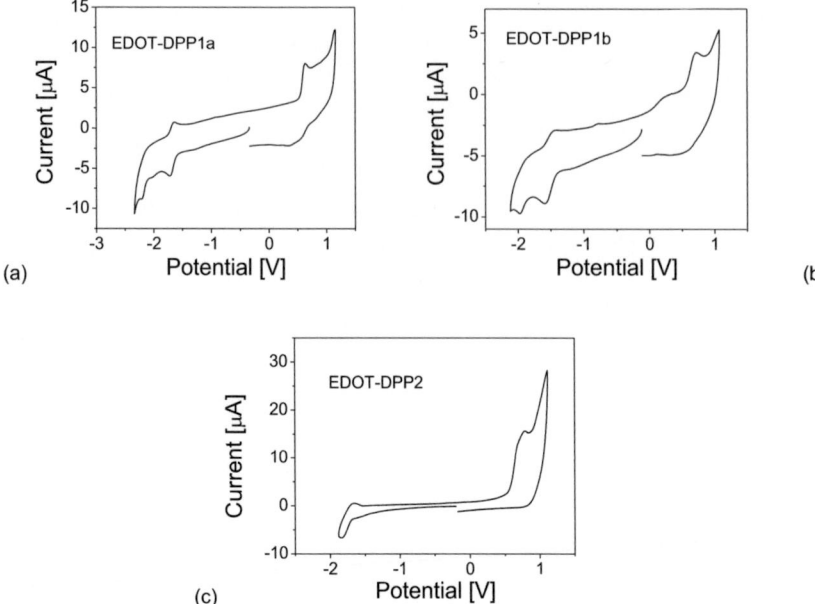

Fig. 3.4: Cyclic voltammograms of **EDOT-DPP1a**, **1b** and **2**, using a carbon working electrode, potential versus an Ag wire pseudo-reference electrode. Supporting electrolyte: 0.1 M TBAPF$_6$ in DCM. Scan rate: 100 mV s^{-1}; T = 20 °C.

Electropolymerization

The electrochemical polymerization of the three EDOT-substituted monomers was studied under potentiodynamic conditions, i.e., the potential was repeatedly cycled between about 0.3 and +0.9 V versus ferrocene, and the current was continuously measured. The plots of electropolymerization of films are displayed in Fig. 3.5. By cycling of **EDOT-DPP1a**, new peaks appeared at +0.21 and +0.45 V, whilst for **EDOT-DPP1b** new peaks at +0.35 and +0.50 V were obtained. The oxidation potentials are lower compared to the corresponding monomers, which is typical for the growth and the elongation of conjugated polymer main chains in **P-EDOT-DPP1a** and **P-EDOT-DPP1b**. Totally different behaviour was shown during the polymerization of **EDOT-DPP2**. The oxidation potential at +0.83 V was shifted to higher values and only weak new peaks appeared in the lower voltage area. The origin for this behaviour is an electrochemical dimerisation of EDOT units of **EDOT-DPP2**, which, however, does not lead to a conjugated polymer chain, but instead only a non-conjugated, insulating chain with isolated diphenyl-substituted bis-EDOT units and DPP units is formed, the π-conjugation being interrupted at the lactam N-atoms.

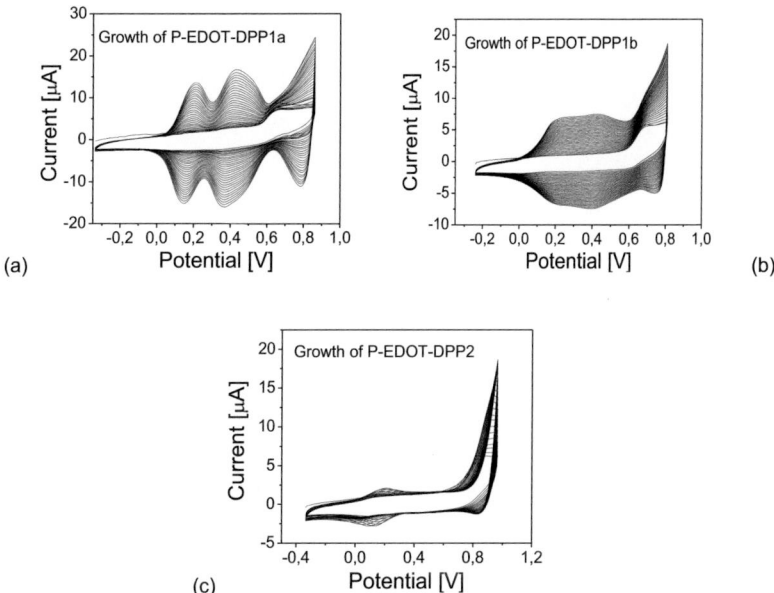

Fig. 3.5: Repeated scan electropolymerization of films of **P-EDOT-DPP1a** (a), **1b** (b) and **2** (c) using a carbon working electrode, potential versus Ag wire pseudo-reference electrode. Supporting electrolyte: 0.1 M TBAPF$_6$ in DCM. Scan rate: 100 mV s^{-1}; T = 20 °C.

It was difficult to further analyse the structure of **P-EDOT-DPP2**. The polymer was quite insoluble. The infrared spectra of **EDOT-DPP2** and **P-EDOT-DPP2** (Fig. 3.6) are well-resolved and rather similar. They especially show the C=O stretching mode at 1680 cm^{-1} and the C=C stretching mode at 1610 cm^{-1}.

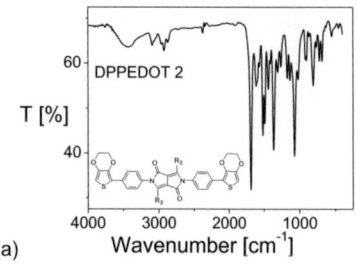 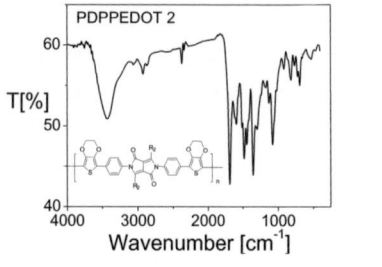

Fig. 3.6: FT-IR spectra of **EDOT-DPP2** (a) and **P-EDOT-DPP2** (b).

The similarity of monomer and polymer spectra can be taken as an indication that the heterocyclic structure of the monomer is retained during polymerization.

Electrochemistry of the polymers

The full redox behaviour of the polymers was also studied using cyclic voltammetry. As shown in Fig. 3.7, the polymers **P-EDOT-DPP1a** and **P-EDOT-DPP1b** displayed reversible oxidation and reduction processes. **P-EDOT-DPP1a** showed reversible oxidation and reduction waves with mid potentials at +0.28 V, +0.54 V and -1.59 V, and **P-EDOT-DPP1b** showed similar behaviour at +0.21 V, +0.50 V and -1.52 V. **P-EDOT-DPP2**, however, only displayed irreversible waves for both processes. The oxidation wave emerged at +1.61 V. It represents the typical oxidation of the non-substituted DPP unit with a potential of about 1.60 V. The polymer chains can be described as an assembly of isolated DPP units, which are connected with each other via EDOT dimer bridges. Along the polymer main chain, the π-conjugation of the DPP units is interrupted at the nitrogen atoms of the lactam groups. The reduction cycle also showed an irreversible process.

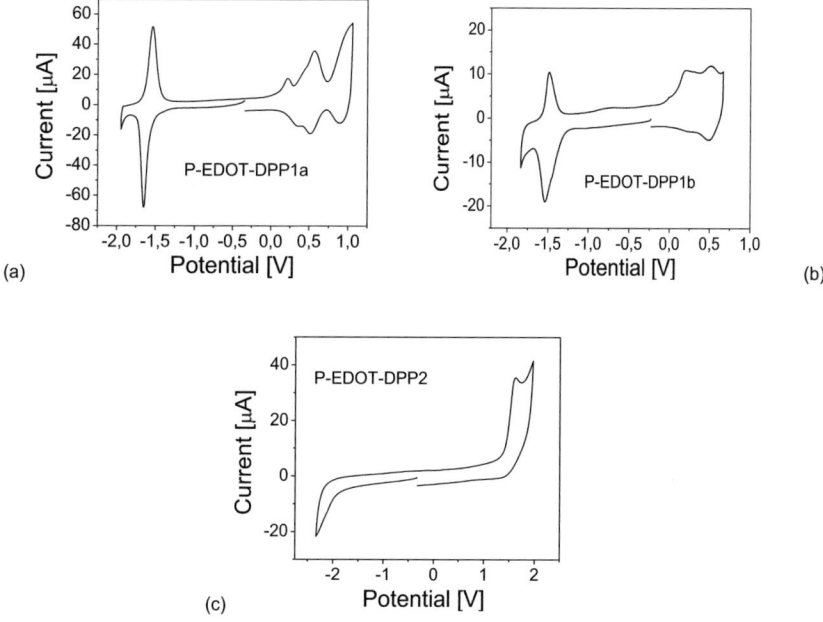

Fig. 3.7: Cyclic voltammograms of **P-EDOT-DPP1a** (a), **P-EDOT-DPP1b** (b) and **P-EDOT-DPP2** (c) as thin films electrodeposited on a glassy carbon electrode, potential versus Ag wire pseudo-reference electrode. Supporting electrolyte: 0.1 M TBAPF$_6$ in DCM. Scan rate: 100 mV s^{-1}; T = 20 °C.

Optical and spectroelectrochemical properties

The new polymers were investigated using UV-Vis spectroscopy and spectroelectrochemistry. Photographs of the films are shown in Fig. 3.8. Compared with the monomers **EDOT-DPP1a** and **EDOT-DPP1b**, the UV/Vis absorption peaks of **P-EDOT-DPP1a**, and **P-EDOT-DPP1b** are shifted to longer wavelengths due to the elongation of the π-conjugated chain (Tab. 3.1).

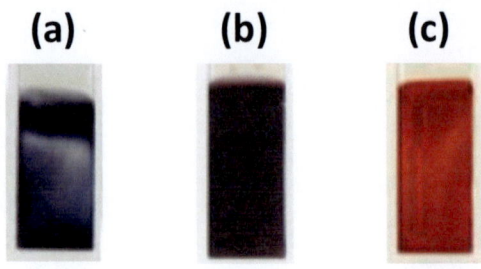

Fig. 3.8. Polymer films of **P-EDOT-DPP 1a** (a), **1b** (b) and **2** (c) in neutral state.

Tab. 3.1: Optical and electrochemical data EDOT-DPP-based monomers and polymers. All redox potentials are referenced to the ferrocene/ferrocenium redox couple.[177]

Compound	UV/Vis λ_{max} / nm	FL λ_{max} / nm	E^1 / V	E^2 / V	E^3 / V	E^4 / V	band gap / eV	oxidation onset / V {HOMO / eV}	reduction onset / V {LUMO / eV}
EDOT-DPP1a	500	581	+0.64	-	-	-1.71	2.13	+0.53 {-5.33}	-1.60 {-3.20}
EDOT-DPP1b	532	585	+0.72	-	-	-1.59	1.93	+0.51 {-5.31}	-1.42 {-3.38}
EDOT-DPP2	492	533	+0.83	-	-	-1.85	2.20	+0.56 {-5.36}	-1.64 {-3.16}
P-EDOT-DPP1a	588	-	+0.28	+0.54	+0.97	-1.59	1.65	+0.08 {-4.88}	-1.57 {-3.23}
P-EDOT-DPP1b	648	-	+0.21	+0.50	-	-1.52	1.37	+0.05 {-4.85}	-1.32 {-3.48}
P-EDOT-DPP2	510	-	+1.61	-	-	-	3.32	+1.41 {-6.21}	-1.91 {-2.89}

E^1/V, E^2/V, E^3/V: first, second and third half-wave oxidation potentials. E^4/V: first reduction potential. Absorption and emission data: monomers in DCM, polymers in solid state.

However, the bathochromic shift of 116 nm for **P-EDOT-DPP1b** is larger than the shift of 88 nm for **P-EDOT-DPP1a**. A possible origin is that the conjugated polymer chain of **P-EDOT-DPP1a** is less planar than **P-EDOT-DPP1b** due to the presence of the bulky hexyldecyl substituent groups. In contrast, **P-EDOT-DPP2** with its absorption maximum at 492 nm exhibits almost no bathochromic shift. The band gap values of **P-EDOT-DPP1a** and **P-EDOT-DPP1b** are 1.65 eV and 1.37 eV, respectively, which are distinctly smaller compared to **EDOT-DPP1a** with 2.13 eV, and **EDOT-DPP1b** with 1.93 eV. **P-EDOT-DPP2**, consequently, exhibits a larger band gap of 3.32 eV, which is also larger than that of **EDOT-DPP2** with 2.20 eV (see Tab. 3.1).

Electrochromism

By oxidation, the UV/Vis absorption spectra of the polymer films underwent a further bathochromic shift. During potential cycling, the colour of the films of **P-EDOT-DPP1a** and **P-EDOT-DPP1b** repeatedly switched between deep blue, transparent grey and purple red. The electrochromism is displayed in Fig. 3.9.

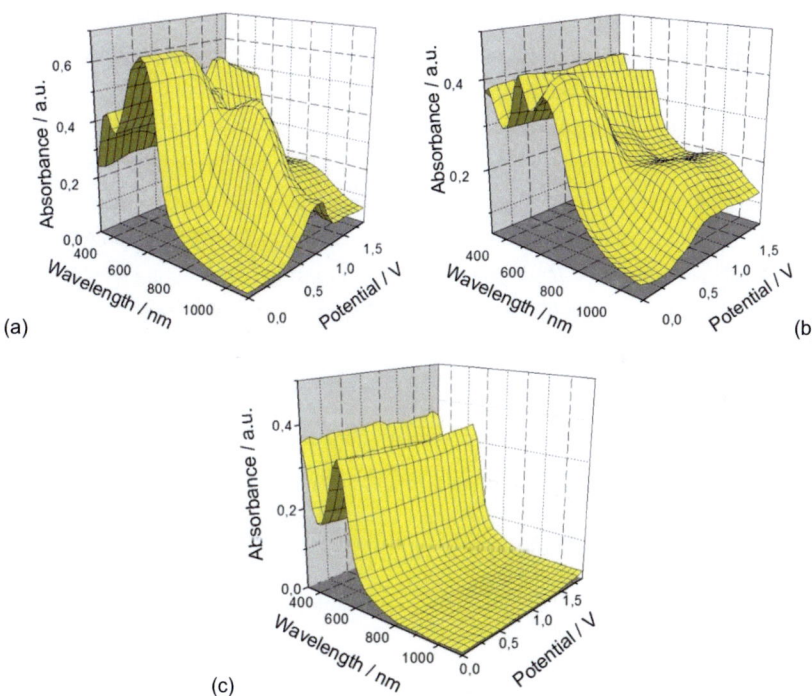

Fig. 3.9: Absorption spectroelectrochemical plots for **P-EDOT-DPP1a** (a), **1b** (b) and **2** (c) as thin films on ITO glass. Solvent: 0.1 M TBAPF$_6$ in acetonitrile. Potential calculated versus ferrocene. Scan rate: 100 mV s^{-1}; T = 20 °C.

With increasing potential the maximum absorption of **P-EDOT-DPP1a** at about 588 nm begins to diminish and a new band with maximum at 785 nm appear. The same process is also observed for **P-EDOT-DPP1b**, the absorption maxima are switched between 648 nm and 893 nm. **P-EDOT-DPP2** exhibits an absorption maximum at about 510 nm, which is not changed by oxidation. For this polymer, an optical band gap of 1.99 eV was calculated, whereas the electrochemical band gap is 3.32 eV, a really high value.

Summary

Within the series of the polymers **P-EDOT-DPP1a**, **1b** and **2**, the optical and electronic properties are strongly influenced by the substitution pattern of the DPP units in the polymer main chain. The influence of the presence of either N-alkyl or N-aryl groups in the DPP units was found to only have a small effect on the optical and electronic properties of the monomers and polymers, whereas the influence of the EDOT-addition to the phenyl rings in 3- and 6-, or 2- and 5- position was very strong. With regard to **EDOT-DPP1a** and **1b**, the π-system is extended by electropolymerization, and **P-EDOT-DPP1a** and **1b** show typical characteristics for conjugated polymers (bathochromic shifts for π-π* transitions, lower oxidation and reduction potentials, narrower band gaps). For **P-EDOT-DPP2**, however, there is no bathochromic shift by electropolymerization, its oxidation potential is higher than for **EDOT-DPP2**, and the oxidation is irreversible. The reason is the interruption of the π-conjugation at the lactam N-atoms causing the polymer chains to consist of electronically isolated DPP-units and isolated phenyl-EDOT-EDOT-phenyl units.

It should be pointed out that due to their narrow band gaps and the reversibility of oxidation and reduction processes, **P-EDOT-DPP1a** and **1b** might be useful for electronic device applications.

3.2 2,3,5,6-Tetra-substituted pyrrolo[3,4-c]pyrrole-1,4(2H,5H)-dione (DPP)-based conjugated polymers

3.2.1 2,3,5,6-Tetra-functionalized pyrrolo[3,4-c]pyrrole-1,4(2H,5H)-dione (DPP) monomers

In Chapter 3.1 it has been shown that the incorporation of a 2,3,5,6-tetraarylated DPP-chromophore in a conjugated linear polymer chain via its 2- and 5- aryl units (i.e., the N-aryl units) results in non-π-conjugated polymers, whereas incorporation via the aryl units in the 3- and 6-positions results in π–conjugated polymers with a bathochromic shift of the optical absorption by about 120 nm (Fig. 3.10).

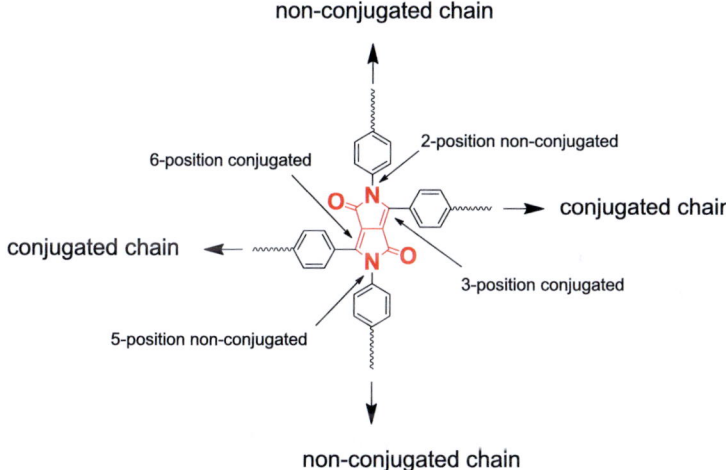

Fig. 3.10: Structure of 2,3,5,6-substituted DPP.

Therefore, it was of great interest to study the electropolymerization of DPP monomers with electropolymerizable units such as thiophene derivatives attached to all the four phenyl groups in the 2-, 3-, 5- and 6-position. In this chapter, the electropolymerization of a series of tetra-functionalized DPP monomers will be described. In three monomers, 3-hexylthiophene-, 3,4-ethylenedioxythiophene- (EDOT-) and 3,4-ethylenedithiathiophene- (EDTT-) groups were attached to the phenyl groups in the 2-, 3-, 5-, and 6-positions of the DPP core (Fig. 3.11). The mechanism of the polymer growth will be studied. Special emphasis is paid to differences of the coupling patterns between the 2,5- and 3,6-positions, and the optical and electrochemical properties of the cross-linked and the linear DPP polymers.

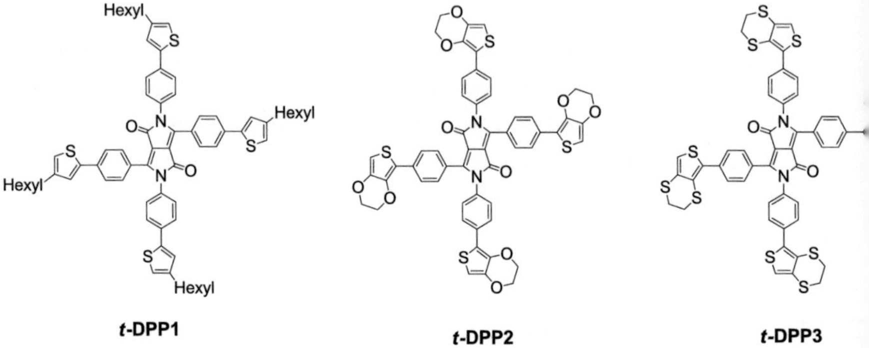

Fig. 3.11: Structures of monomers *t*-DPP1 - 3.

Synthesis

The synthetic route to the key compound 2,3,5,6-tetrakis(4-bromophenyl)pyrrolo[3,4-c]-pyrrole-1,4(2H,5H)-dione (**t-BrDPP**) and the compounds **t-DPP1 - 3** is described in Scheme 3.3. The starting compound **BrDFF** was synthesized as described in Chapter 3.1.1.

The synthesis of **t-BrDPP** required a condensation reaction of **BrDFF** with *p*-bromoaniline, in which the lactone units were converted into lactam units. A brilliant red solid product was obtained in 55% yield. The 3-hexyl-thiophene-, 3,4-ethylenedioxythiophene- (EDOT-) and 3,4-ethylenedithia- thiophene- (EDTT-) substituted monomers **t-DPP1 - 3** were synthesized using microwave assisted Stille coupling of monomer **t-BrDPP** with 4-hexyl-thien-2-yl trimethylstannane (**HTH1**), 3,4-ethylenedioxythien-2-yl trimethylstannane (**EDOT1**) and 3,4-ethylenedithiathien-2-yl trimethyl- stannane (**EDTT1**) using Pd(PPh$_3$)$_4$ as the catalyst and DMF as the solvent. The microwave assisted coupling methods gave high yields of 86, 71 and 81% for *t*-**DPP1, 2** and **3**, respectively. All the three monomers are red solids, which are soluble in common organic solvents such as chloroform, DCM, DMF, THF, and toluene, for example.

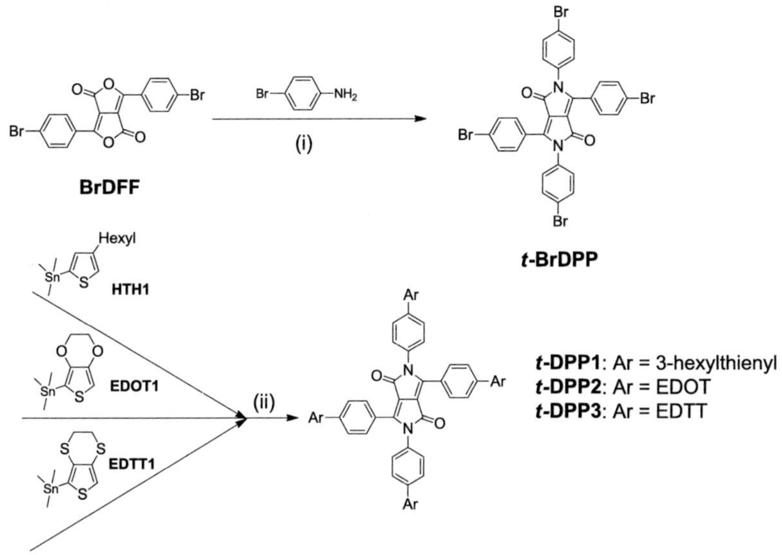

Scheme 3.3. Synthetic route to the tetra-functional DPP-monomers. Reagents and conditions: (i) *N,N'*-dicyclohexylcarbodiimide, CF_3COOH, $CHCl_3$, 3d. (ii) $Pd(PPh_3)_4$, DMF, microwave, 1h.

The ^{1}H NMR spectra of *t*-**DPP1 - 3** displayed all the expected resonances with no discernible peaks corresponding to impurities. Significantly, the singlet signal at around 6.9 ppm can be ascribed to the thiophene unit in the 3-hexylthiophene-, EDOT- and EDTT- substituents. The two triplet signals with a chemical shift of about 4.34 ppm are typical for the ethylene bridge of the EDOT unit. The two triplet signals with a chemical shift of about 3.37 ppm are typical for the ethylene bridge of the EDTT unit with reference to the literature.[174, 178]

The UV/vis absorption and fluorescence emission spectra of the key compound *t*-**BrDPP** are shown in Fig. 3.12. *t*-**BrDPP** shows absorption maxima at 470 and 503 nm, and an emission maximum at 543 nm. The spectra are similar to these of the bifunctional monomers **BrDPP1** and **2** (see Chapter 3.1).

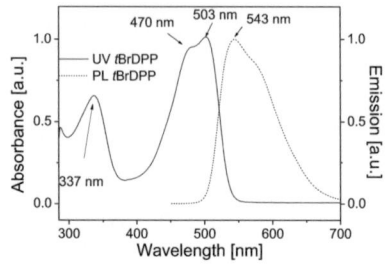

Fig. 3.12. UV/vis absorption and fluorescence spectra of *t*-BrDPP.

UV/vis absorption of the monomers

The UV/vis absorption spectra of **t-DPP1 - 3** are similar (Fig. 3.13). Compared to the starting compound **t-BrDPP** (λ_{max} = 502 nm), bathochromic shifts are observed by the addition of the thiophene derivatives to the conjugated system. While the EDOT-substituted monomer **t-DPP2** showed a large shift of 33 nm, the 4-hexylthienyl- and EDTT-substituted monomers **t-DPP1** and **t-DPP3** exhibit smaller shifts of 18 and 17 nm, respectively. The reason could be the larger electron withdrawing effect caused by the sulfur in the EDTT unit compared to oxygen in the EDOT unit, which leads to a less extended conjugated system in **t-DPP3**. As for **t-DPP1**, 3-hexylthiophene is not as electron-rich as EDOT, which leads to a less extended conjugated system compared with **t-DPP2** with EDOT-substituents. The onset absorption edges of the monomers occur at about 580 nm, corresponding to optical HOMO-LUMO gaps in the range from 2.11 to 2.17 eV. Significantly, the extinction coefficients of the monomers are very high (Tab. 3.2).

Tab. 3.2. Extinction coefficients of *t*-DPP1 - 3.[a]

Monomer	Absorption λ_{max} / nm	Extinction coefficients ε / L mol^{-1}cm^{-1}
t-DPP1	520	$7.3 \cdot 10^3$
t-DPP2	535	$7.7 \cdot 10^3$
t-DPP3	519	$5.3 \cdot 10^3$

[a] Absorption spectra were taken in DCM solution.

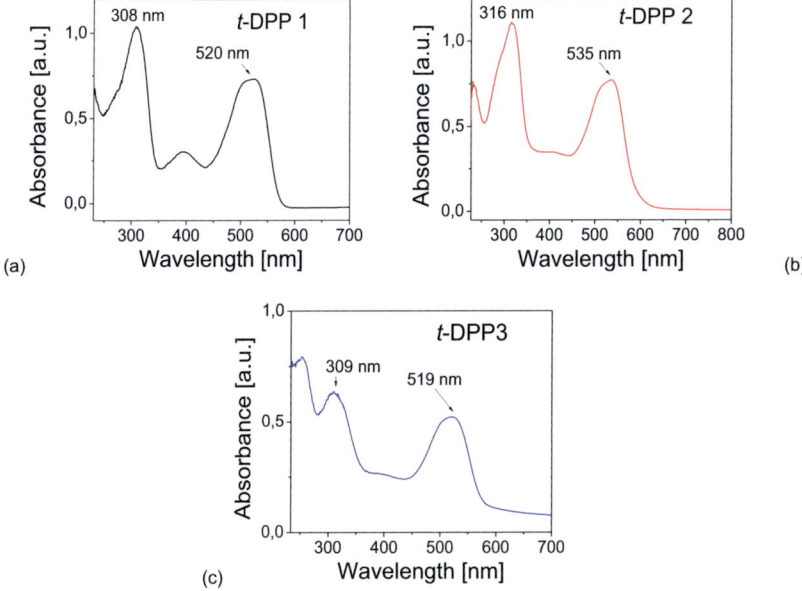

Fig. 3.13. UV/vis absorption spectra of monomers *t*-DPP1 - 3.

Fig. 3.14. Photographs of monomers **BrDFF**, *t*-**BrDPP** and *t*-**DPP1** - **3** in DCM solution.

Electrochemistry

The electrochemistry of **t-DPP1**, **t-DPP2**, and **t-DPP3** was studied using cyclic voltammetry in DCM solution. Glassy carbon was used as working electrode, a platinum coil as counter electrode, Ag/AgCl as reference electrode, and tetrabutylammonium hexafluorophosphate as the supporting electrolyte. The electropolymerization was carried out under potentiodynamic conditions. The cyclic voltammograms are displayed in Fig. 3.15. The monomers exhibit quasi-reversible anodic waves at about +0.80 V and cathodic waves at about -1.45 V, which are reverted at -1.35 V. The electrochemical HOMO-LUMO gaps are 2.07 eV for **t-DPP1**, 1.98 eV for **t-DPP2**, and 2.00 eV for **t-DPP3**, which are in good agreement with the optically determined HOMO-LUMO gaps. Anodic oxidation of **t-DPP1** is indicated by two quasi-reversible peaks at +0.78 and +0.91 V, which could be caused by the two different thiophene-aryl-substituents in the 2,5- and 3,6-positions of the DPP core. **t-DPP2** and **t-DPP3**, however, only showed one oxidation peak at about +0.70 V. However, the broad shape of the peak suggests that it represents a superposition of two very close peaks of the EDOT-aryl- and EDTT-aryl-substituents in either the 2,5- or 3,6-position of the central DPP chromophore. The data are collected in Tab. 3.3.

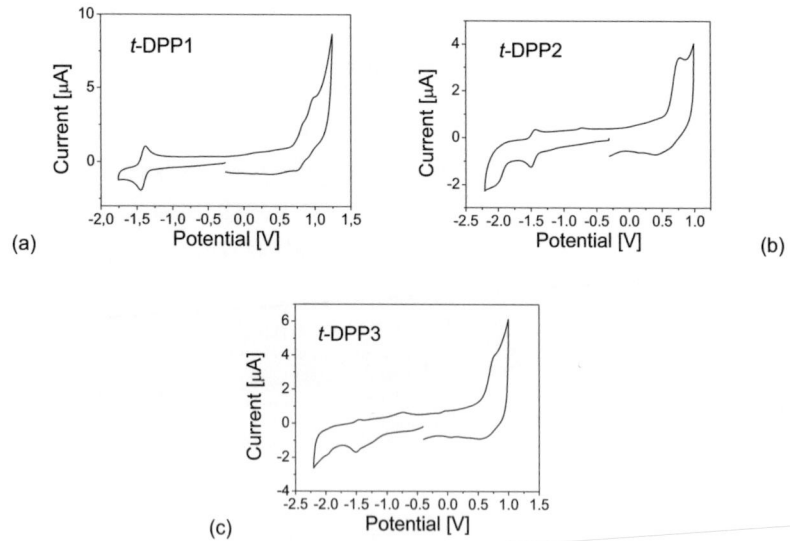

Fig. 3.15. Cyclic voltammograms of monomers **t-DPP1 - 3** in solution. Solvent: 0.1 M TBAPF$_6$ in DCM. Potential calculated versus ferrocene. Scan rate: 100 mV s^{-1}; T = 20 °C.

Tab. 3.3. Optical and electrochemical data of monomers[a] t-DPP1 - 3.

	UV/nm	HOMO-LUMO gap (opt) / eV	Onset of oxidation / V	Onset of reduction / V	HOMO {LUMO} / eV	HOMO-LUMO gap / eV
t-DPP1	309, 395, 520	2.17	+0.71	-1.36	-5.51{-3.44}	2.07
t-DPP2	317, 535	2.11	+0.59	-1.39	-5.39{-3.41}	1.98
t-DPP3	253, 309, 519	2.14	+0.59	-1.41	-5.39{-3.39}	2.00

[a]Absorption spectra were taken in DCM solution. All redox potentials refer to the ferrocene/ferrocenium redox couple.

3.2.2 Electropolymerization, optical and electrochemical properties of the polymers

The three monomers were electropolymerized using the same solvent and conditions as for the cyclic voltammetric experiments, i.e., the potential was cycled repetitively over the redox active range of the materials from -0.25 to +1 V for **P-t-DPP1**, and from -0.30 to +0.75 V for **P-t-DPP2** and **P-t-DPP3**. The polymer growth plots are displayed in Fig. 3.16.

t-DPP2 and **t-DPP3** polymerized readily, while **t-DPP1** required twice as many cycles, which could be caused by the high oxidation potential of thiophene without electron donating groups as in EDOT or EDTT,[179] or as the consequence of slow coupling kinetics for oxidised **t-DPP1**. During the polymer growth many different signals appeared. This could be in accordance with the random growth of the polymers, which is obtained through all thiophene units (top/bottom to left/right), and not only through those in the 3- and 6-positions (left to right) or 2- and 5-positions (top to bottom). According to the study of electropolymerization of bisEDOT-DPP derivatives in Chapter 3.2.1,[100] a random growth could be likely because the thiophene units in the 3- (left) and 6- (right) positions are conjugated to each other meaning that any radical cation formed is stabilised and resonated over this conjugation before radical combination could allow for polymer growth.

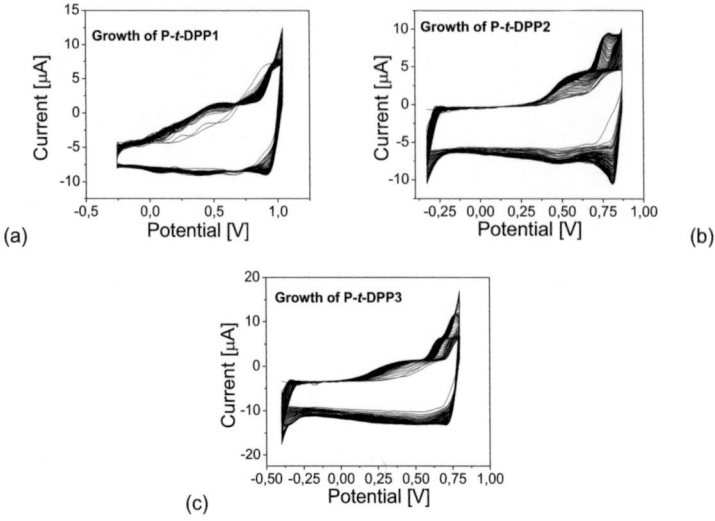

Fig. 3.16. Repeated scan electropolymerization of **P-*t*-DPP1** (a), **P-*t*-DPP2** (b), and **P-*t*-DPP3** (c) using a carbon working electrode, and Ag wire pseudo-reference electrode. Supporting electrolyte: 0.1 M TBAPF$_6$ in DCM. Scan rate: 500 mV s^{-1}; T = 20 °C.

For the three compounds, new peaks emerged at lower potentials, which have its origin in the growth and elongation of conjugated polymer main chains. The polymers exhibit HOMO-LUMO gaps in a range from 1.68 to 1.80 eV, which are lower than for the insulating polymer only grown through the 2,5-positions (E_g > 3 eV) (see Chapter 3.2.1).[180] Polymers obtained from the monomers *t*-DPP1 to 3 probably contain short conjugated chains, in which the conjugation is frequently interrupted due to a random reaction of the monomer units in the 2,3,5, and 6-position.

The probabilities of formation of differently sized conjugated blocks are plotted in Fig. 3.17. The 'degree of polymerization' indicates the number of monomer units being connected to form the polymer chain, e.g., a degree of three indicates that the polymer chain consists of three monomer units. The term 'number of conjugated units' indicates, how many of the monomer units in the polymer chain are in π-conjugation with each other.

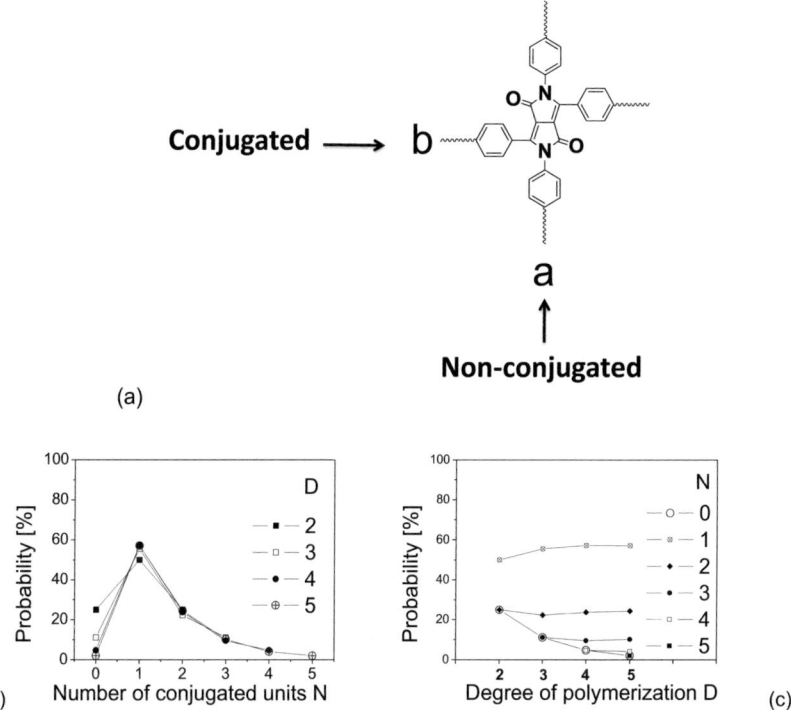

Fig. 3.17. Probabilities study for differently sized conjugated blocks. D: Degree of polymerization; N: number of conjugated units in polymer chain, for N = 0, the polymer chain is fully non-conjugated. For N = 5, the polymer chain is fully conjugated.

For a chain with degree of polymerization of 2, four possibilities exist: The two monomers can be coupled via their b-positions (see Fig 3.17a), a-positions, a- and b-, or b- and a-position. In the first case a conjugated dimer is obtained, the number n of conjugated units is 2. In the second case, a fully non-conjugated dimer is formed, n is zero. In the third and fourth case one of the two units is conjugated and the other one is not, n is 1. For n = 2 and n = 0, the probabilities are 25% each, for n = 1 it is 50%. In Fig. 6b the probability is plotted versus the number n of conjugated units for different degrees of polymerization. It is shown, for example that in a polymer with degree of polymerization of five the presence of an isolated conjugated unit (n=1) exhibits 58% probability, whereas the existence of conjugated dimers, trimers, and tetramers only has 25, 12, and 6% probability, respectively. The formation of a fully conjugated or non-conjugated chain is the least probable scenario (Fig.

3.17b). Fig. 3.17c indicates that there is little change in the probabilities of differently sized conjugated blocks with increasing degree of polymerization.

Cyclic voltammograms of polymers

The electrochemistry of the electrochemically grown polymer films was investigated in acetonitrile vs. Ag/AgCl. The cyclic voltammograms of **P-*t*-DPP1**, **P-*t*-DPP2**, and **P-*t*-DPP3** are shown in Fig. 3.18. **P-*t*-DPP1** exhibits irreversible oxidation and reduction waves, from which a HOMO-LUMO gap of 1.20 eV can be calculated. The diagram of **P-*t*-DPP2** is complex. The oxidation first shows a small and broad peak at low potential (-0.12 V), which could be from bis-EDOT units in the 3- and 6-positions (left to right) of the chain, and the wave at higher potential (+0.54 V) indicates the oxidation of the EDOT units in the 2- and 5-positions (top to bottom). It is possible that the large reduction peak at (-0.70 V) is the reverse of the oxidation process, but the peak itself shows some reversibility, which makes the assignment of this peak ambiguous.

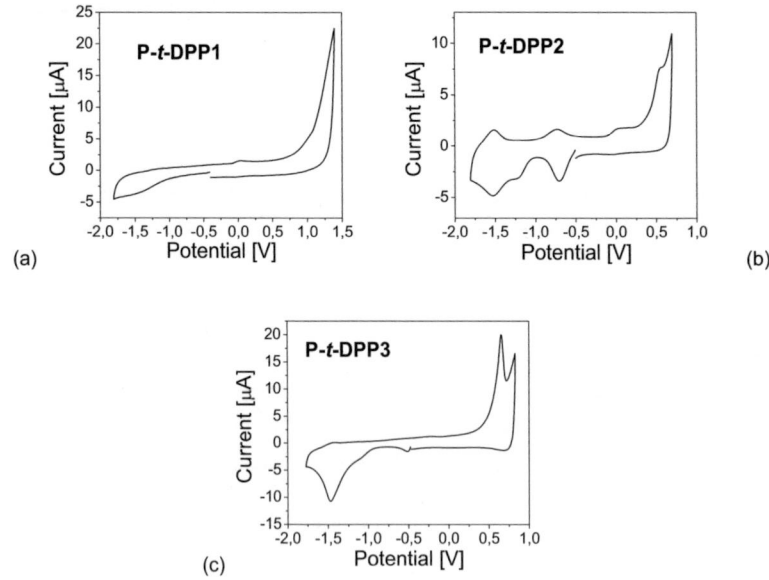

Fig. 3.18. Cyclic voltammograms of polymers **P-*t*-DPP1 - 3** as thin films deposited on a glassy carbon electrode. Solvent: 0.1 M TBAPF$_6$ in acetonitrile. Potential calculated versus ferrocene. Scan rate: 100 mV s^{-1}; T = 20 °C.

The reduction of **P-*t*-DPP2** includes a cathodic wave at -1.53 V, which is reverted at -1.51 V and is based on the same process of the DPP core seen in the monomer reduction (***t*-DPP2**). There is also a shoulder at -1.30 V, which could be the reduction of a short conjugated chain. A HOMO-LUMO gap of 1.23 eV was determined. **P-*t*-DPP3** produced a less complex cycle. By anodic oxidation, a large irreversible wave emerged at +0.66 V, and by reduction, an irreversible wave appeared at -1.45 V. A much smaller redox couple occured at -0.28 / -0.51 V, which could indicate the same process as shown for **P-*t*-DPP2**, in which the conjugated units in the 3- and 6-positions are oxidized. The other difference between the two polymers is that the small peak at +0.01 V in **P-*t*-DPP2** (possibly being due to bis-EDOT) is absent in **P-*t*-DPP3**. This may indicate poor conjugation between two EDTT units, which are twisted in relation to each other and therefore represent poor electron donors.[173] Finally, the reduction of the DPP core is irreversible in the case of **P-*t*-DPP3**. The HOMO-LUMO gap was determined as being 1.48 eV.

UV/vis spectroscopic and spectroelectrochemical studies

It is of interest to study the spectroelectrochemical properties of this series of polymers. Thin polymer films were grown on ITO and electronic absorption spectra were taken. The 3-D electronic absorption plots for **P-*t*-DPP1**, **P-*t*-DPP2** and **P-*t*-DPP3** are shown in Fig 4.19, the optical data are listed in Tab. 3.4. In addition to the UV/vis absorption maxima at 504 nm for **P-*t*-DPP1**, 510 nm for **P-*t*-DPP2**, and 511 nm for **P-*t*-DPP3**, the three polymers showed broad shoulders from about 600 to 800 nm indicating an elongation of the $\pi-\pi^*$-conjugated system along the polymer main chain. The spectroelectrochemistry of **P-*t*-DPP1** showed very little change until a potential of +1.3 V was reached. At this potential the intensity of the $\pi-\pi^*$ transition began to decrease, and the broad absorption shoulder starting from 650 nm increased. In contrast, **P-*t*-DPP2** showed no change in the absorption characteristics with increasing potential. It could be possible that the polymer produced was highly cross-linked and, due to S-O interactions adjacent monomer units were densely packed so that counter ions could not diffuse into the film to balance the charge and allow any change to occur (Fig. 3.20). [181]

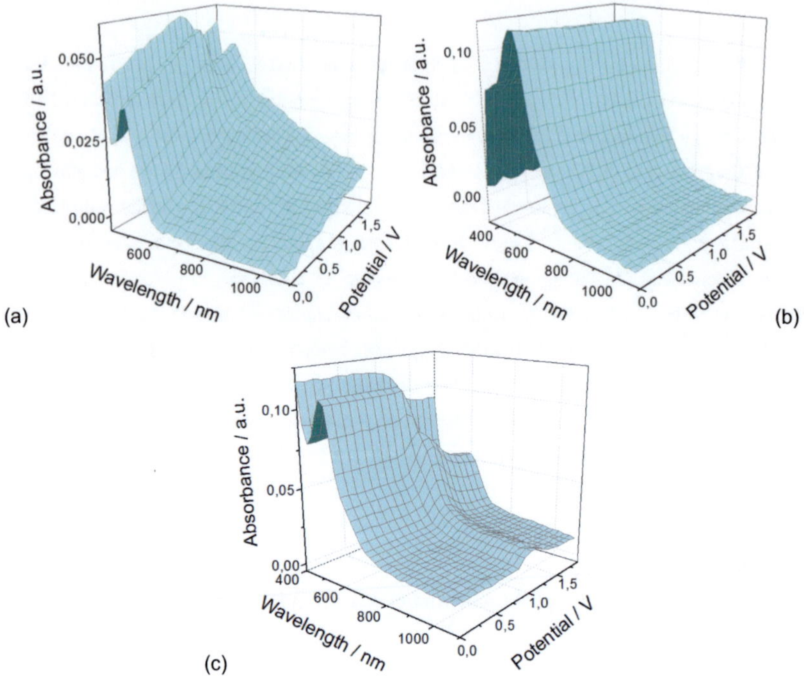

(a)

(b)

(c)

Fig. 3.19. Absorption spectroelectrochemical plots for **P-*t*-DPP1** (a), **P-*t*-DPP2** (b), and **P-*t*-DPP3** (c) as thin films on ITO. Ag wire pseudo-reference electrode. Solvent: 0.1 M TBAPF$_6$ in acetonitrile. Potential calculated versus ferrocene. Scan rate: 100 mV s^{-1}; T = 20 °C

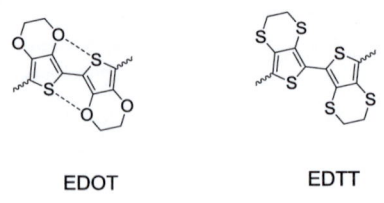

EDOT   EDTT

Fig. 3.20. Sulfur oxygen interaction between adjacent EDOT and EDTT units.

Tab. 3.4. Optical and Electrochemical Data of Polymers[a] P-t-DPP1 - 3.

	UV/nm	HOMO-LUMO gap (opt) / eV	Onset of oxidation / V	Onset of reduction / V	HOMO{LUMO} / eV	HOMO-LUMO gap / eV
P-t-DPP1	342, 504	2.07	-0.09	-1.01	-4.71{-3.79}	1.20
P-t-DPP2	510	1.72	-0.12	-1.35	-4.68{-3.45}	1.23
P-t-DPP3	511	1.70	-0.25	-1.23	-4.55{-3.57}	1.48

[a]Absorption spectra were taken from films. All redox potentials refer to the ferrocene/ferrocenium redox couple.

The spectroelectrochemistry experiment for **P-t-DPP3** showed significant changes in the electronic absorption spectra with increasing potential. At +1.0 V, one can see the formation of polarons and bipolarons in the polymer chain which results in a new peak between 500 and 700 nm, followed by a drop in the peak at 510 nm. Between +1.0 and +1.4 V there was a weak broad peak appearing from 700 nm to the infrared. The S-O interaction between adjacent EDOT-units in **P-t-DPP2** does not exist in EDTT-containing **P-t-DPP3**. So the polymer is less likely to be densely packed, allowing easier access for the counterions to diffuse into the polymer to balance the charge.

The reasoning for the large discrepancy between the optically and electrochemically determined band gaps is supported by the spectroelectrochemical results. For all three polymers, there is relatively little change in the absorption spectra and this indicates that oxidation takes place at localized, short-conjugation sites. The LUMO of the materials is therefore dominated by the DPP core and the HOMO is derived from the thiophene segments. The absorption spectra provide information about the π-π^* transitions and the corresponding orbitals are clearly not representative of the HOMO and LUMO orbitals (since the gaps determined by the two methods differ by 0.2-0.9 eV). In conjugated polymers, this situation often arises, if there is a strong redox-active unit in the structure, which is poorly conjugated to the main chain.[182]

As an interesting comparison, the linear polymers containing 2,5-bis(4´-t-butylphenyl)-3,6-diphenylpyrrolo[3,4-c]pyrrole-1,4(2H,5H)-dione (see Chapter 3.2.1)[180] with EDOT only attached to the 3,6-positions showed fully reversible oxidative and reductive waves. The polymer was fully conjugated and electrochromic, giving a HOMO-LUMO gap of 1.32 eV. The other polymer containing DPP units with EDOT only attached to the 2,5-positions showed irreversible oxidative and reductive waves, giving a HOMO-LUMO gap of 3 eV. This polymer was neither conjugated nor electrochromic. The observed behaviour suggests that the three polymers **P-*t*-DPP1**, **P-*t*-DPP2** and **P-*t*-DPP3** are not fully conjugated, because cross-linking via the 2,3,5, and 6-directions is equally likely (although only cross-linking via the 3- and 6-positions leads to extension of the conjugated system).

In summary, a series of electropolymerizable tetra-functionalized DPP monomers have been synthesized, and the corresponding polymers have been prepared by anodic electropolymerization. The study shows that the polymer growth takes place randomly through the 2-, 3-, 5-, and 6-position of the DPP core. Compared with the monomers, **P-*t*-DPP1**, **P-*t*-DPP2** and **P-*t*-DPP3** showed broad long wavelength absorption bands, which indicates an enlargement of the π–conjugation in the polymers. However, compared with linear polymers only grown through the 3,6-positions, the tetrafunctionalized monomers lead to polymers with less π–conjugation and only little pronounced electrochromic properties.

In polymers **P-*t*-DPP1**, **P-*t*-DPP2** and **P-*t*-DPP3**, conjugated blocks through coupling of thiophene units in the 3,6-positions were obtained, which are separated by non-conjugated blocks. As a result, the polymers are not as fully conjugated as the linear polymers prepared previously through coupling in the 3,6-positions.

3.3 Conjugated microporous poly(pyrrolo[3,4-*c*]pyrrole-1,4(2H,5H)-dione) (DPP) networks

In Chapters 3.2.1 and 3.2.2 we studied the incorporation of pyrrolo[3,4-*c*]pyrrole-1,4(2H,5H)-dione (DPP) into conjugated linear or cross-linked polymers prepared electrochemically under anodic conditions. It will be of considerable interest to study the same cross-linked polymers prepared by conventional methods such as metal-catalyzed polycondensation reactions. In this chapter, we will study cross-linked polymers based on pyrrolo[3,4-*c*]pyrrole-1,4-dione (DPP) using microwave assisted nickel-mediated or palladium-catalyzed condensation reactions, such as Yamamoto-[68] or Sonogashira-Hagihara-[75] cross-coupling polycondensation. The polymers were characterized by elementary analysis, N_2 gas sorption, FT-IR spectroscopy and SEM. The synthetic route is described in Scheme 3.4.

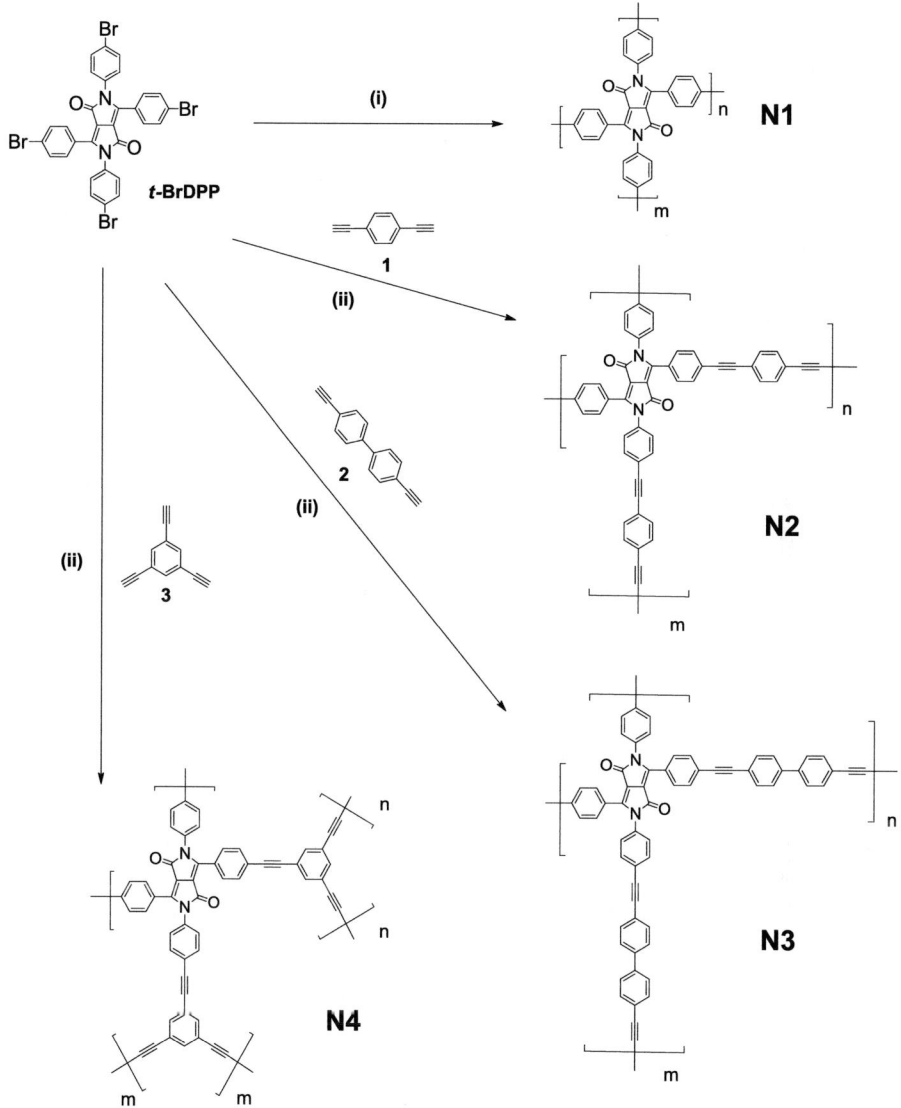

Scheme 3.4. Idealized structures of the polymer networks **N1 - 4**. Reagents and conditions: (i) Ni(COD)$_2$, 2,2'-dipyridyl, cyclooctadiene, DMF, microwave, 100 °C, 1h. (ii) Pd(PPh$_3$)$_4$, NEt$_3$, DMF, microwave, 100 °C, 1h.

The self-condensation of 2,3,5,6-tetrakis(4-bromophenyl)pyrrolo[3,4-c]pyrrole-1,4(2H,5H)-dione *t*-BrDPP using Ni(COD)$_2$, 2,2'-dipyridyl and cyclooctadiene gave the polymer **N1**. Polymers **N2**, **N3**, and **N4** were synthesized by palladium-catalyzed Sonogashira-Hagihara-cross-coupling polycondensation of monomer *t*-BrDPP and arylethynylenes.

All the reactions were carried out under microwave assisted conditions in a short reaction time (1 h). The polymers precipitated as red powders and were totally insoluble in all solvents used (Fig. 3.21).

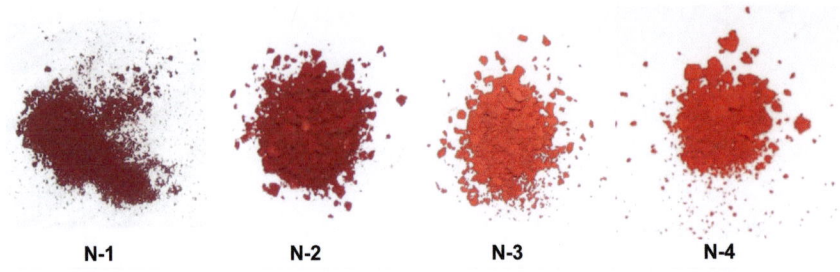

Fig. 3.21. Photographs of the insoluble polymer networks **N1 - 4**.

The FT-IR spectra of the polymers showed stretching modes similar to those of the starting compound *t*-BrDPP. From 2900 to 3400 cm^{-1} the C-H stretching vibration appeared. The typical C=O stretching mode of the DPP chromophore appeared at about 1690 cm^{-1} and the C=C stretching mode at 1600 cm^{-1}. Additionally, polymers **N2**, **N3** and **N4** showed the C≡C stretching mode at about 2200 cm^{-1}. The FT-IR spectra of the starting compound *t*-BrDPP and the polymers **N1 - 4** are shown in Fig. 3.22.

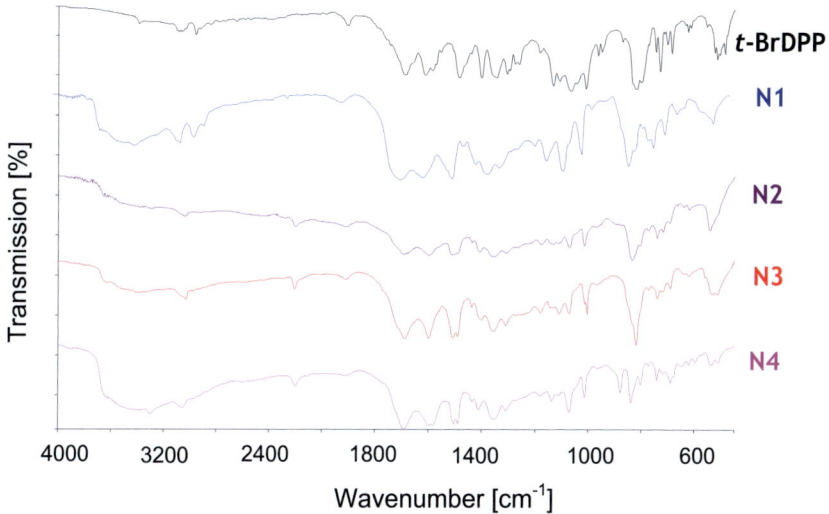

Fig. 3.22. FT-IR spectra of *t*-BrDPP and **N1 - 4**.

High resolution magic angle spinning NMR

^{1}H-NMR spectra of insoluble materials can be taken using the high resolution magic angle spinning NMR method. The spectra display similar proton signals to standard ^{1}H-NMR spectra taken in solvents. Similar to the starting compound *t*-BrDPP, the polymer networks display typical signals of aromatic protons of the phenyl rings attached to the DPP units, and for **N2 - 3** also the aromatic protons of the comonomers appear in the range from 6.8 to 8.0 ppm.

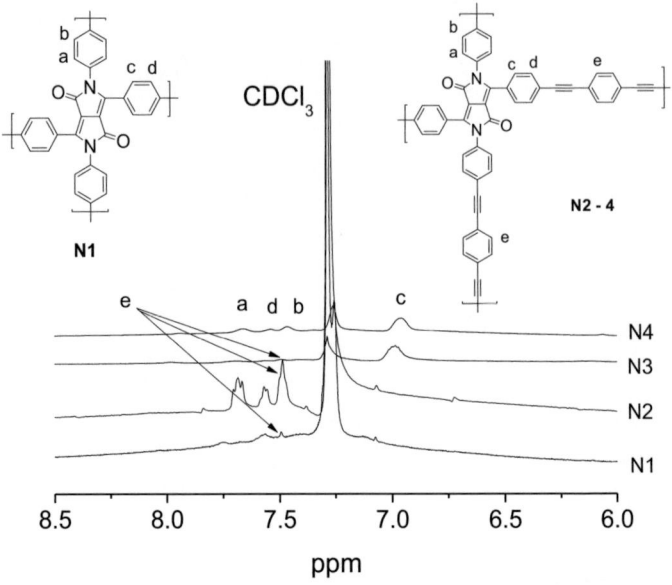

Fig. 3.23. ¹H-NMR spectra of polymer networks **N1 - 4** taken in CDCl₃ using the high resolution magic angle spinning NMR method.

The broad signals of the chemical shifts at about 7.69 ppm can be ascribed to the aromatic protons in the **a** positions of the 2,5-substitutated phenyl rings (Fig. 3.23). The broad signals between 7.56 and 7.49 ppm can be ascribed to protons in the **b** and **d** positions of the phenyl rings. Similar to **t-BrDPP**, the protons in the **c** positions of the 3,6-substituted phenyl rings show very broad signals with the chemical shifts at in the range from 6.90 to 7.10 ppm. As for the networks **N2 - 4**, the aromatic protons of the comonomers are displayed as broad signals with chemical shifts at around 7.50 ppm, the shifts being the same as for the other proton signals of the phenylene units of DPP.

Surface areas and pore size distributions were measured by nitrogen adsorption and desorption at 77.3 K. The nitrogen isotherms are shown in Fig. 3.24.

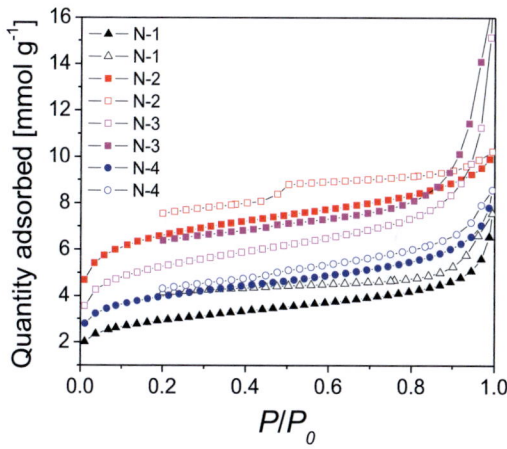

Fig. 3.24. Nitrogen sorption isotherms for poly-DPP networks **N 1 - 4**. Adsorption branches are labeled with filled symbols.

Using the analytic methods as described in Chapter 2.7, the polymer networks were investigated as the following. The BET (Brunauer, Emmett, and Teller)[155] surface areas were calculated in the range from 216 to 500 m² g^{-1}. According to IUPAC classifications, the adsorption isotherms suggest that the polymers are microporous structured networks. Polymer **N2** shows the largest BET surface area of 476 m² g^{-1} among the polymer networks, with a slightly N_2 hysteresis appearing in the desorption branch. The pore size distribution curves derived from the Barret-Joyner-Halenda (BJH) method[183] are displayed in Fig. 3.25. The total pore volumes at a relative pressure of P/P_0 = 0.99 were calculated in the range from 0.23 to 0.52 m³ g^{-1}. The micropore volumes were calculated in the range from 0.05 to 0.13 m³ g^{-1} using the *t*-plot method[171].

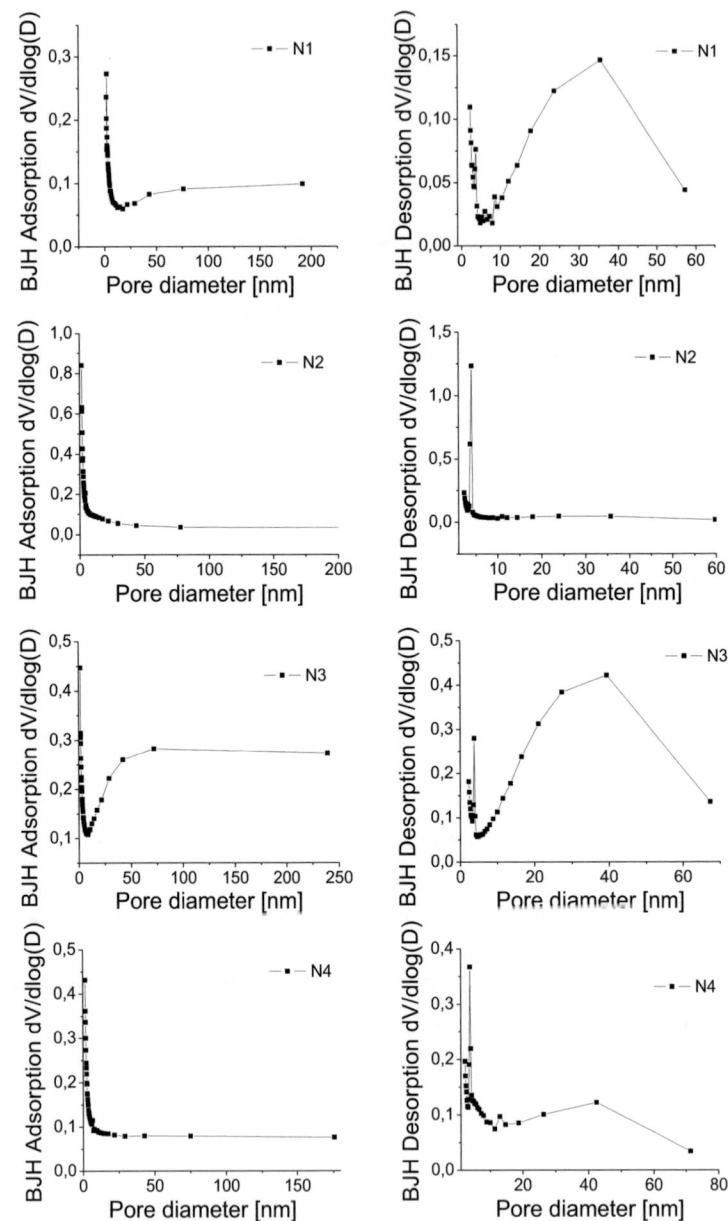

Fig. 3.25. Pore size distributions of networks **N1 - 4**.

Tab. 3.5. Structural properties of networks N1 - 4.

	M1	M2	S_{BET} [m^2g^{-1}][a]	$S_{Langmuir}$ [m^2g^{-1}][b]	Pore diameter/ adsorption [nm]	Pore diameter/ desorption [nm]	$V_{pore/tot}$ [m^3g^{-1}][c]	$V_{pore/micro}$ [m^3g^{-1}][d]	Microporosity [%]
N1	(structure)	--	216	329	6.48	12.08	0.23	0.05	22
N2	(structure)	(structure)	477	713	3.48	4.91	0.34	0.13	38
N3	(structure)	(structure)	384	582	8.71	14.52	0.52	0.09	17
N4	(structure)	(structure)	290	439	5.00	7.33	0.28	0.07	25

[a] Surface areas calculated from the N_2 adsorption isotherms using BET method. [b] Surface areas calculated from the N_2 adsorption isotherms using Langmuir method[164]. [c] Total pore volumes at P/P_0=0.99. [d] Micropore volumes derived using the t-plot method based on the Halsey equation[171].

The micoporosities of the networks are in the range from 17 to 38% with network **N2** showing the highest porosity (Tab. 3.5). The study indicates the microporous nature of the networks. Based on the assumption that the N_2 gas was only adsorbed in monolayers, the surface areas were calculated using the Langmuir method. The data are listed in Tab. 3.5.

SEM analysis

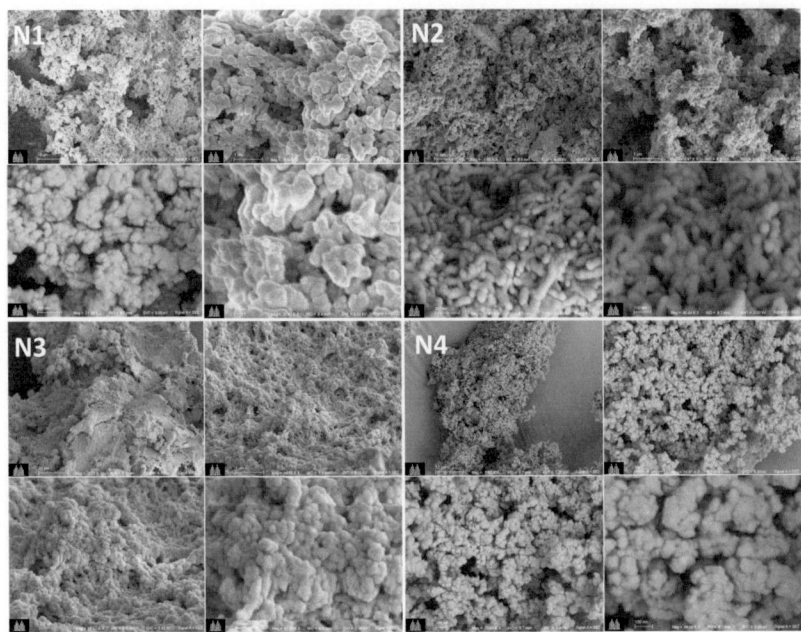

Fig. 3.26. SEM images of the polymer networks (Scale: top left: 10 μm, top right: 1 μm; bottom left: 200 nm; bottom right: 100 nm).

The SEM images of the polymer networks clearly show a very porous surface.

In summary, for the first time we have introduced pyrrolo[3,4-c] pyrrole-1,4(2H,5H)-dione (DPP) into conjugated microporous networks with BET surface areas from 210 - 477 $m^2 g^{-1}$ using nickel- or palladium-catalyzed cross-coupling reactions. The microwave assisted polycondensation reactions offered an alternative preparation method for conjugated polymer networks. It would be of great interest to prepare thin films of these networks and study their electrochemical properties.

3.4 Oligomers based on 2,3,5,6-tetraphenylpyrrolo[3,4-c]pyrrole-1,4(2H,5H)-dione (DPP)

The synthetic route is outlined in Scheme 3.5. Starting from 3,6-diphenylfuro[3,4-c]furan-1,4-dione (DFF) and comonomers *p*-phenylenediamine (**1**), 3,3'-dimethoxy-[1,1'-biphenyl]-4,4'-diamine (**2**), 4,4'-methylenedianiline (**3**), or dodecane-1,12-diamine (**4**), a series of oiligomers was synthesized via simple polycondensation reactions. The yields were in the range from 50 to 60%. In the reaction, oxygen atoms of the lacton groups of DFF were replaced by nitrogen atoms forming lactam groups. As a result, an oligomeric chain was built through the repeating DPP units.

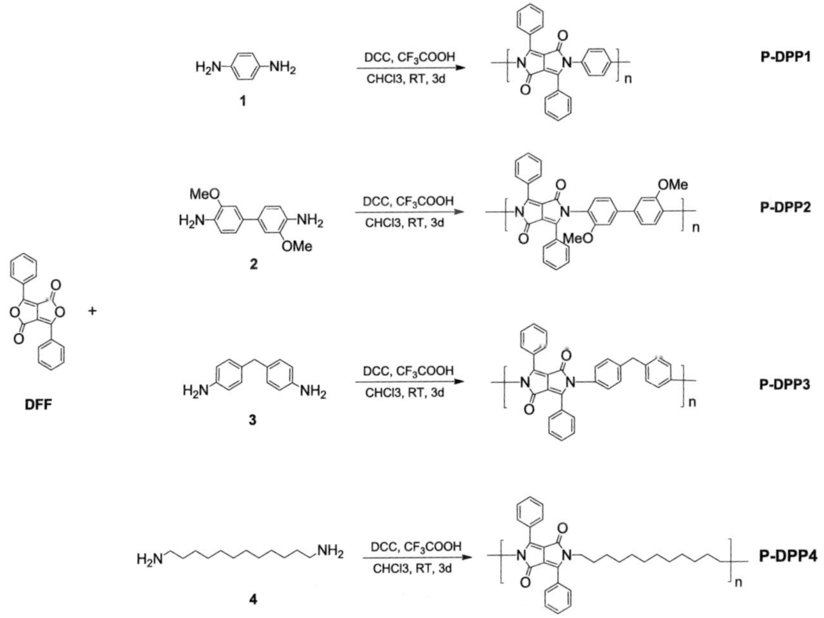

Scheme 3.5. Synthetic route to DPP oligomers **P-DPP1 - 4**.

All the four oligomers are orange solids showing similar molecular weights in the range from 3500 to 4500 Dalton. They are soluble in common organic solvents. In solution they show yellow orange colours. The ^{1}H-NMR spectra (see Chapter 5, Experimental Part) of **P-DPP1 - 4** show the typical broad signals of aromatic protons at 7.8 - 6.9 ppm. For **P-DPP1**, no other proton signals were determined. For **P-DPP2**, broad signals at about 3.9 ppm were shown, which can be ascribed to proton signals of the methoxy groups. For **P-DPP3**, signals at ca. 4.0 ppm can be ascribed to protons of the methylene units. **P-DPP4** shows typical broad signals of the α-CH$_2$ protons next to the lactam nitrogen atom, and the proton signals of the β-CH$_2$ groups. In addition, the other adjacent CH$_2$-proton signals are displayed at about 1.7 ppm.

UV/vis absorption spectra

The UV/vis absorption spectra of **DFF** and **P-DPP1 - 4** are displayed in Fig. 3.27 and 4.28. The optical data are listed in Tab. 3.6. The starting compound **DFF** shows very sharp absorption bands with maxima at 408, 434 and 463 nm (Fig. 3.27). The DCM solution has a greenish colour. From the absorption edge at 477 nm an optical HOMO-LUMO gap of 2.60 eV can be calculated.

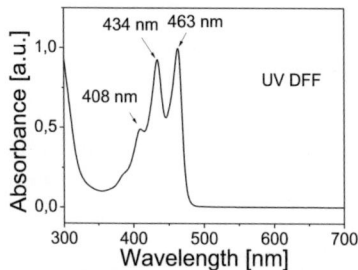

Fig. 3.27. UV/vis absorption spectrum of DFF in DCM.

In comparison, the oligomers **P-DPP1 - 4** show much broader absorption bands with maxima in a range of 450 to 490 nm. In addition they all show broad absorptions in the range from 300 to 400 nm, which can be referred to the typical absorption range of phenyl groups. Compared to the starting compound **DFF**, **P-DPP3** and **4** exhibit bathochromic shifts of about 30 nm. which is caused by the replacement of oxygen atoms in the lacton rings by nitrogen atoms. The conjugated system expands due to the additional π-electrons of the nitrogen atoms in the DPP units. **P-DPP1** and **2** show hardly any bathochromic shifts compared to **DFF**. However, they exhibit broad absorption shoulders between 500 and 550 nm, indicating an extension of the conjugated system. After the addition of the two comonomers *p*-phenylendiamine and 3,3'-dimethoxy-[1,1'-biphenyl]-4,4'-diamine the oligomer chains can be twisted and lack planarity, which could favour absorption maxima in lower wavelength regions.

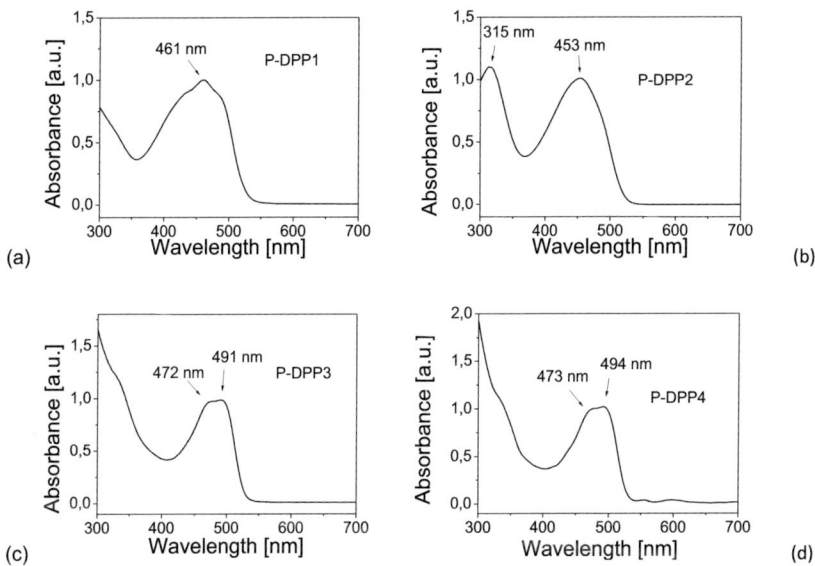

Fig. 3.28. UV/vis absorption spectra of **P-DPP1 - 4** in DCM solution.

Electrochemistry

The electrochemistry of the oligomers was investigated under the same potentiodynamic conditions as described in Chapter 3.1. The cyclic voltammograms are displayed in Fig. 3.29 and the electrochemical data listed in Tab. 3.6. All the four oligomers show quasi-reversible oxidative and reductive waves. **P-DPP1** shows a quasi-reversible oxidative wave at +0.74 and an irreversible wave at +1.15 V, the reductive cycle shows a cathodic wave at -1.31, which is reverted at -1.18 V. **P-DPP2** only displays irreversible oxidative and reductive waves at +1.04 and -1.26 V, respectively. The cycle of **P-DPP3** is similar to **P-DPP1**. It shows quasi-reversible oxidative waves at +0.37 and +0.80 V, and reversible reductive waves at -1.07 and

−1.38 V. However, the oxidation and reduction processes of **P-DPP4** are irreversible. The HOMO-LUMO gaps of the series of oligomers vary in the range from 1.30 to 1.90 eV.

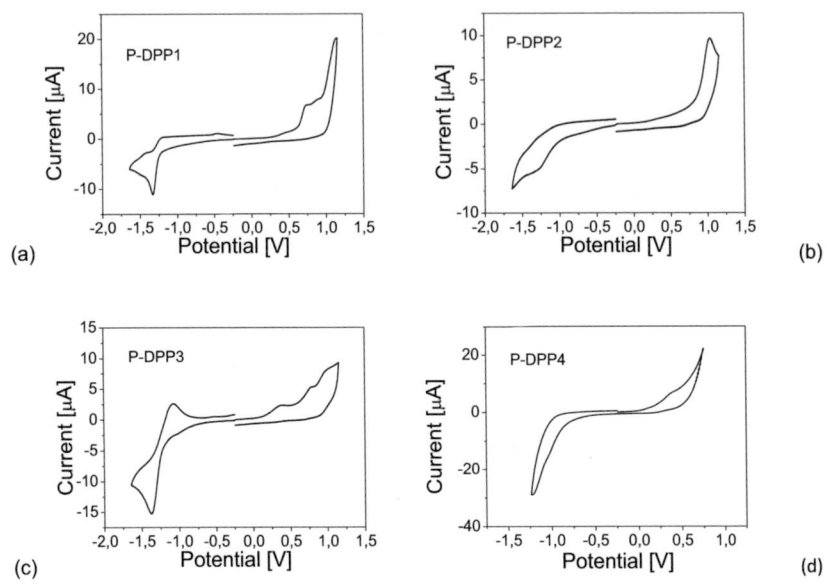

Fig. 3.29. Cyclic voltammograms of polymers **P-DPP1 - 4** as thin films deposited on a glassy carbon electrode. Solvent: 0.1 M TBAPF$_6$ in acetonitrile. Potential calculated versus ferrocene. Scan rate: 100 mV s^{-1}; T = 20 °C.

Tab. 3.6. Optical and electrochemical data of DFF and P-DPP1 - 4.[a]

	UV / nm	HOMO-LUMO gap (opt) / eV	Onset of oxidation / V	Onset of reduction / V	HOMO {LUMO} / eV	HOMO-LUMO gap / eV
DFF	408, 434, 463	2.60	-	-	-	-
P-DPP1	461	2.37	+0.63	−1.23	−5.43{−3.57}	1.86
P-DPP2	315, 453	2.40	+0.88	−1.03	−5.68{−3.77}	1.91
P-DPP3	472, 491	2.36	+0.45	−1.18	−5.25{−3.62}	1.63
P-DPP4	473, 494	2.35	+0.40	−0.92	−5.20{−3.88}	1.32

[a]Absorption spectra were taken in dicloromethane solution. All redox potentials refer to the ferrocene/ferrocenium redox couple.

In summary, a series of oligomers based on 2,3,5,6-tetraphenylpyrrolo[3,4-c]pyrrole-1,4(2H,5H)-dione were synthesized using simple polycondensation reactions. In fact, the optical and electrochemical properties of the oligomers are similar to isolated DPP units. The reason is that the oligomer chains are not π-conjugated, the conjugation is interrupted at the nitrogen atoms of the lactam rings of the DPP core. This leads to almost the same optical and electrochemical HOMO-LUMO gaps for the oligomers and individual DPP units.

3.5 Conjugated polymers based on thienyl-DPP

Conjugated polymers and block copolymers were synthesized via palladium-catalyzed Suzuki-coupling polycondensation reactions. Special emphasis was paid to the influence of the *t*-butoxycarbonyl (**Boc**) substituent of the DPP core on the optical and electrochemical properties of the polymers.

3.5.1 Conjugated alternating polymer containing Boc-substituted thienyl-DPP

The first sample of Boc-substituted thienylDPP-based polymer was synthesized and characterized using UV/vis absorption spectroscopy, cyclic voltammetry and thermal analysis. The synthetic route to the starting compound **BrBoc-DPP**, and polymer **P-Boc-ThDPP1** is described in Scheme 3.6.

The acylation of the thienylDPP pigment **ThDPP** was obtained in a high yield of 89%. The monomers **Boc-ThDPP** and **BrBoc-ThDPP** are dark red solids. Compared to the pigment **ThDPP**, they show a very high solublity in common organic solvents. The ^{1}H-NMR spectra of **Boc-ThDPP** and **BrBoc-ThDPP** displayed all the typical proton signals as expected. The signals at 8.24, 7.65, and 7.20 ppm can be ascribed to the protons of the thiophene units and the very intensive peak at 1.64 ppm is the CH_3-proton signal of the Boc-groups. The polymer **P-Boc-ThDPP1** was prepared via Suzuki coupling polycondensation reaction in a yield of 82% using thiophene diboronic ester as the comonomer. **P-Boc-ThDPP1** is a dark solid showing a deep purple colour in solution.

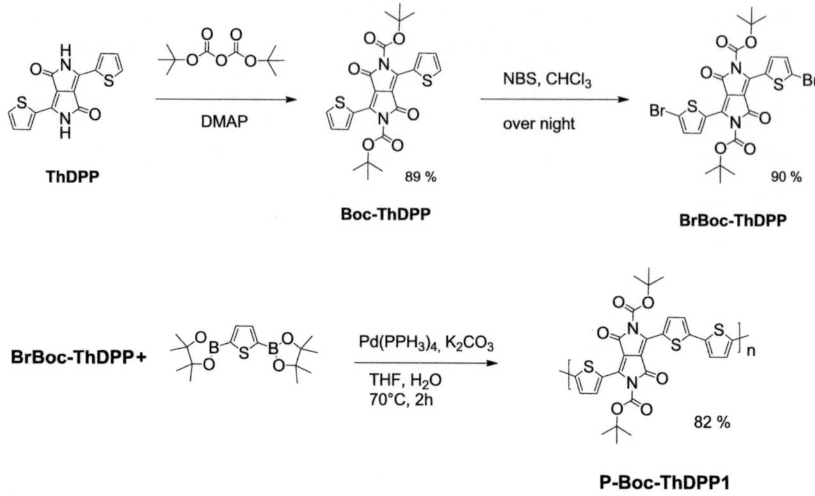

Scheme 3.6. Synthetic route to monomer **BrBoc-ThDPP**, and polymer **P-Boc-ThDPP1**.

UV/vis absorption spectroscopy

The UV/vis absorption spectra of the monomers and polymer are shown in Fig. 3.30, and their photographs in DCM solution are displayed in Fig. 3.31. **ThDPP** shows absorption maxima at 533 and 495 nm. The acylated monomers **Boc-ThDPP** and **BrBoc-ThDPP**, however, exhibit absorptions with maxima at 495 nm for **Boc-ThDPP** and 515 nm for **BrBoc-ThDPP**, respectively. The reason is that for **ThDPP**, the strong hydrogen bonding between the lactam units and short distances between the chromophore planes (0.336 nm) and phenyl ring planes (0.354 nm) enable $\pi-\pi$-interactions via molecular orbital overlapping and exciton coupling effects, which lead to an ordered planar structure. The acylated monomers **Boc-ThDPP** and **BrBoc-ThDPP** do not possess the strong hydrogen bonding anymore. The lack of the hydrogen bonding provides a much better solubility and weaker $\pi-\pi$-interactions between the molecules. This may explain the hypsochromic shifts in the UV/vis spectra of **Boc-ThDPP** and **BrBoc-DPP** compared with **ThDPP**. The polymers **P-Boc-ThDPP1** shows an absorption maximum at 541 nm, and a very broad shoulder between 560 and 750 nm, which indicates an elongation of the $\pi-\pi^*$-conjugated system in direction of the polymer main chain. From the absorption onset at about 729 nm, an optical HOMO-LUMO gap of 1.7. eV can be calculated.

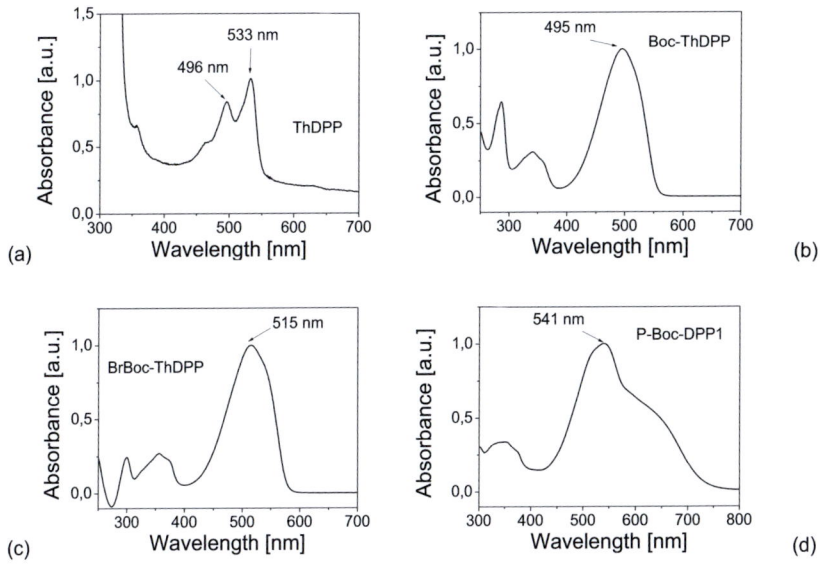

Fig. 3.30. UV/vis absorption spectra of **ThDPP** (a), **Boc-ThDPP** (b) and **BrBoc-ThDPP** (c) and **P-Boc-ThDPP1** (d).

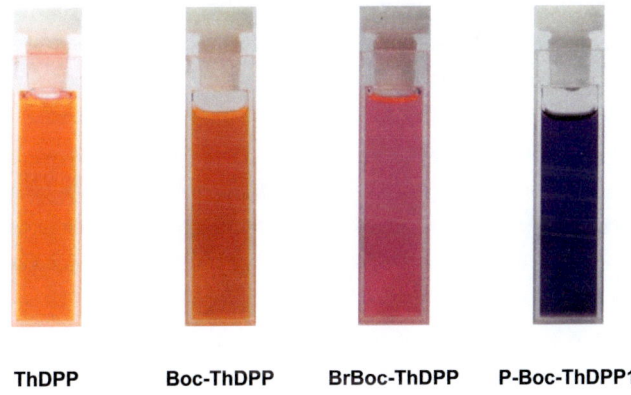

Fig. 3.31. Photographs of **ThDPP** in DMSO solution, **Boc-DPP**, **BrBoc-ThDPP** and **P-Boc-ThDPP1** in DCM solution.

Thermal cleavage of the Boc-group

It is known that the Boc-substituent can be removed via thermal treatment.[184] Therefore, a thin film of polymer **P-Boc-ThDPP1** was thermally treated at 180 °C for different time periods (Scheme 3.7). UV/vis absorption spectra and cyclic voltammograms were taken before and after the thermal treatment. The spectra of the polymer films are displayed in Fig. 3.32, and the optical data are compiled in Tab. 3. 8.

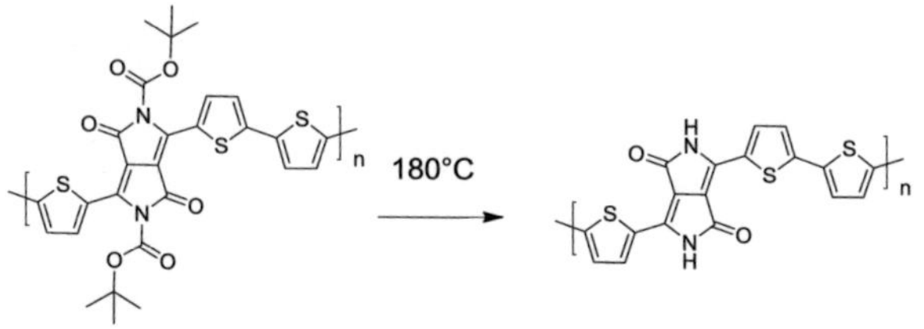

P-Boc-ThDPP1

Scheme 3.7. Thermal cleavage of the Boc-groups of **P-Boc-ThDPP1**.

P-Boc-ThDPP1 exhibits absorption bands with the maximum being at 612 nm in the solid state. It can be shown that after thermal treatment at 180 °C for 30 min, 1 h and 18 h, the absorption spectra of the polymer film do not differ much from each other. This indicates that the Boc-group can be almost completely removed after only a short period of 30 min by thermal treatment at 180 °C. The spectra also indicate that the absorption is not much changed by the thermal treatment, the maximum of 612 nm is maintained. However, the broader shoulders between 700 and 800 nm indicate a better $\pi-\pi^*$-conjugation in direction of the polymer main chain. The reason could be that after removal of the Boc-groups the strong hydrogen bonding between the lactam units is reset and the polymer chains become more planar. This favours a better extension of the $\pi-\pi^*$-conjugated system along the polymer chain.

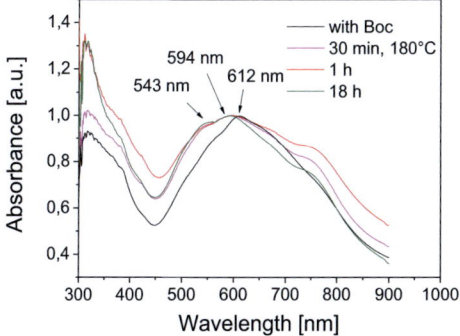

Fig. 3.32. UV/vis absorption spectra of **P-Boc-ThDPP1** as thin film before and after thermal treatment.

IR spectroscopy

The thermal cleavage of the Boc-group in the monomer **BrBoc-ThDPP** and polymer **P-Boc-ThDPP1** was investigated using infrared spectroscopy. The spectra were taken before and after the thermal treatment at 180 °C after a time period of 30 min. The spectra are displayed in Fig. 3.33.

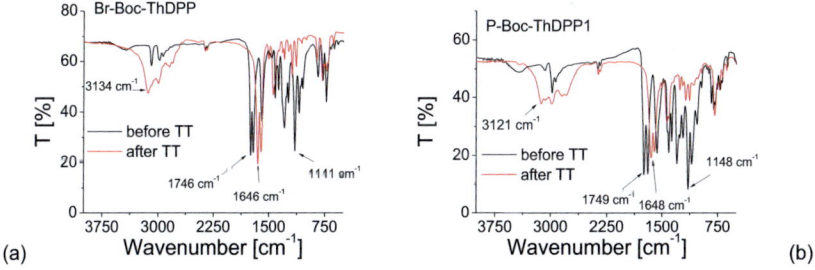

Fig. 3.33. IR spectra of **BrBoc-ThDPP** (a) and **P-Boc-ThDPP1** (b) before (black band) and after (red band) thermal treatment at 180 °C after a time period of 30 min.

The infrared spectra of **BrBoc-ThDPP** and **P-Boc-ThDPP1** show similar bands before and after the thermal treatment. Especially the C = O stretching mode of the Boc-group at 1746 cm^{-1} disappeared after heating at 180 °C for 30 min, only the C = O stretching mode of the lactam group at 1646 cm^{-1} retained. Significantly, the N - H stretching mode at 3130 cm^{-1} appeared, which indicates the reset of the hydrogen bond after the cleavage of the Boc-group. The C(CH$_3$)$_3$ - deformation at about 1145 cm^{-1} decreased largely after the thermal treatment.

Electrochemistry

Before the thermal treatment at 180 °C, **P-Boc-ThDPP1** shows a quasi-reversible anodic wave at +0.88 V, which is reverted at +0.36 V, and a cathodic wave at -1.78, which is reverted at -0.94 V. After the thermal treatment, the polymer shows an anodic wave at +0.62 V, and a cathodic wave at -1.39 V. During the experiment, the polymer film was more stable after thermal treatment compared to the untreated polymer film. After removal of the Boc-group, the polymer shows lower HOMO-LOMO gaps either optically or electrochemically. The optical and electrochemical data are listed in Tab. 3.7.

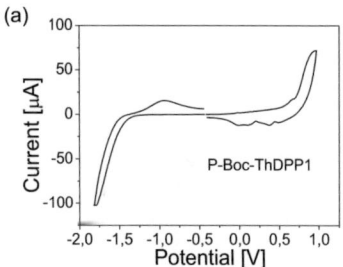 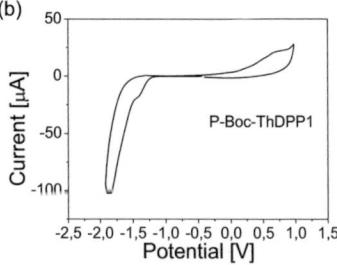

Fig. 3.34. Cyclic voltammograms of **P-Boc-ThDPP1** before (a) and after thermal treatment at 180 °C for 30 min (b). Sample: A thin film deposited on ITO. Solvent: 0.1 M TBAPF$_6$ in acetonitrile. Potential calculated versus ferrocene. Scan rate: 100 mV s^{-1}; T = 20 °C.

Tab. 3.7. Optical and electrochemical data of ThDPP, Boc-ThDPP, BrBoc-ThDPP and P-Boc-ThDPP1.[a]

	UV / nm	HOMO-LUMO gap (opt) / eV	Onset of oxidation / V	Onset of reduction / V	HOMO {LUMO} / eV	HOMO-LUMO gap / eV
ThDPP	496, 533	2.25	-	-	-	-
Boc-ThDPP	495	2.23	-	-	-	-
BrBoc-DPP	515	2.16	-	-	-	-
P-Boc-ThDPP1 before TT	612	1.40	+0.54	-1.48	-5.34{-3.32}	2.02
after TT	594	1.30	+0.20	-1.43	-5.00{-3.37}	1.62

[a]Absorption spectra of monomers were taken in dicloromethane solution, for **P-Boc-ThDPP1** as thin film. All redox potentials refer to the ferrocene/ferrocenium redox couple. TT: thermal treatment.

3.5.2 Conjugated copolymers based on Boc-ThDPP

A series of conjugated copolymers based on Boc-ThDPP was synthesized using palladium-catalyzed Suzuki-coupling polycondensation reaction. They were characterized using UV/vis absorption spectroscopy, cyclic voltammetry and thermal analysis. The synthetic pathway is shown in Scheme 3.8.

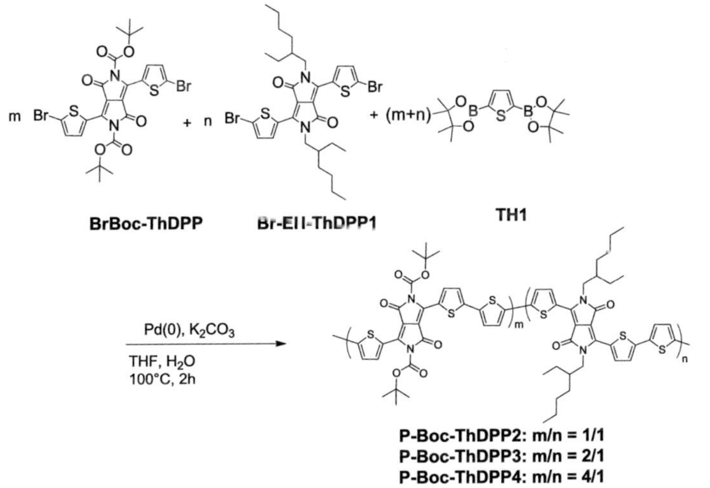

P-Boc-ThDPP2: m/n = 1/1
P-Boc-ThDPP3: m/n = 2/1
P-Boc-ThDPP4: m/n = 4/1

Scheme 3.8. Synthetic route to conjugated block copolymers containing different **Boc-ThDPP** content.

The polymers were obtained in yields of around 75%. They were dark solids. The ^{1}H-NMR spectra of the polymers displayed identical proton signals as **P-Boc-ThDPP1**, except that polymers **P-Boc-ThDPP2 - 4** show additional signals between 2.4 and 0.8 ppm, which can be ascribed to the proton signals of the alkyl chain. The signals at 8.30, 7.60, and 7.20 ppm can be ascribed to the protons of the thiophene units and the very intensive peak at 1.70 ppm is the CH_3-proton signal of the Boc-groups. (see Chapter 5, Experimental Part).

Optical and electrochemical properties before and after the thermal treatment

In the solid state the polymers show a deep red colour with absorption maxima in the range from 600 to 620 nm. The broad absorption shoulders between 700 and 900 nm indicate an elongation of the $\pi-\pi^*$-conjugated system in direction of the polymer chains. The UV/vis spectra of the polymers **P-Boc-ThDPP2 - 4** are displayed in Fig. 3.35. The optical and electrochemical data are compiled in Tab. 3.9. Photographs of the polymers in DCM solution are shown in Fig. 3.36.

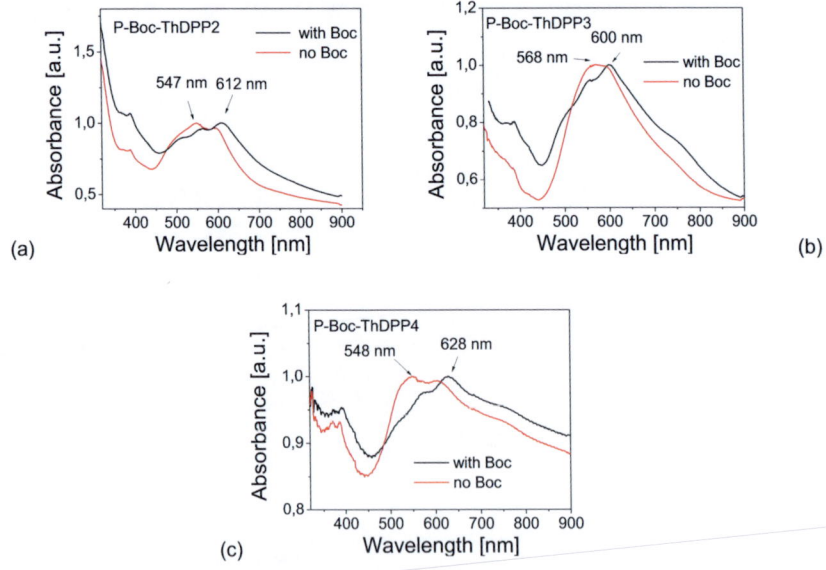

Fig. 3.35. UV/vis absorption spectra of **P-Boc-ThDPP2** (a), **3** (b) and **4** (c) as thin films before and after thermal treatment at 180 °C for 30 min.

The polymers show very broad absorption bands with maxima in the range from 600 to 630 nm, exhibiting broad shoulders between 700 and 900 nm. After the thermal treatment, the polymers exhibit broad absorption bands with maxima in the range from 550 to 570 nm, which are similar to **P-Boc-ThDPP1**. However, the broad shoulders between 700 and 900 nm indicate a longer π–conjugation in direction of the polymer chain.

P-Boc-ThDPP2   P-Boc-ThDPP3   P-Boc-ThDPP4

Fig. 3.36. Photographs of **P-Boc-ThDPP2 - 4** in DCM solution.

IR spectroscopy

The thermal cleavage of the Boc-group of **P-Boc-ThDPP2 - 4** was investigated using infrared spectroscopy. The spectra were taken before and after the thermal treatment at 180 °C after a time period of 30 min. The spectra are displayed in Fig. 3.37. The infrared spectra of the polymers show similar bands before and after the thermal treatment. Similar to **P-Boc-ThDPP1**, the C = O stretching mode of the Boc-group in the three polymers at about 1750 cm^{-1} disappeared after heating at 180 °C for 30 min, only the C = O stretching mode of the lactam group at about 1660 cm^{-1} retained. Significantly, the N   H stretching mode at about 3120 cm^{-1} appeared, which indicates the reset of the hydrogen bond after the cleavage of the Boc-group. The C(CH$_3$)$_3$ - deformation at about 1145 cm^{-1} decreased largely after the thermal treatment.

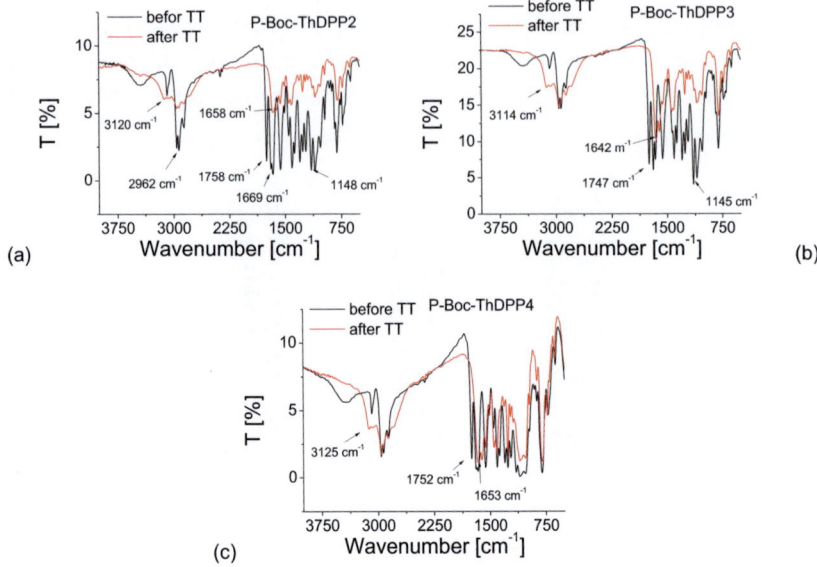

Fig. 3.37. IR spectra of **P-Boc-ThDPP2** (a), **3** (b) and **4** (c) before (black band) and after (red band) thermal treatment at 180 °C after a time period of 30 min.

Electrochemistry

Under identical conditions as for **P-Boc-ThDPP1**, **P-Boc-ThDPP2 - 4** were investigated using cyclic voltammetry before and after a thermal treatment at 180 °C for 30 nm. The cyclic voltammograms of the polymers are shown in Fig. 3.37, and the electrochemical data are listed in Tab. 3.8. **P-Boc-ThDPP2** shows a reversible anodic wave at +0.79 V, which is reverted at +0.51 V, and an irreversible cathodic wave at -1.55 V. After thermal treatment, the polymer shows an anodic wave at +0.76 V, which is reverted at +0.37 V, and an irreversible cathodic wave at -1.83 V. **P-Boc-ThDPP3** exhibits an anodic wave at +1.26 V, which is reverted at +1.09 V, and a cathodic wave at -1.77 V. After thermal treatment, the polymer shows an irreversible oxidative wave at +1.04 V and a cathodic wave at -1.73 V. The oxidation cycle of **P-Boc-ThDPP4** is reversible. It shows an anodic wave at +0.87 V, which is reverted at +0.70 V. The reduction cycle of the polymer appears irreversible. After the thermal treatment, the polymer shows an anodic wave at +0.59 V, which is reverted at +0.47 V, and two cathodic waves at -1.25 and -1.65 V.

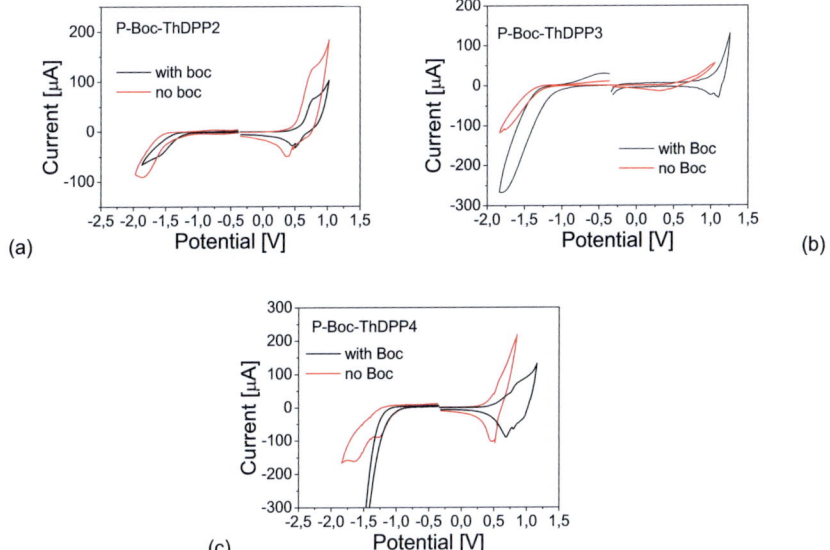

Fig. 3.37. Cyclic voltammograms of **P-Boc-ThDPP2** (a), **3** (b) and **4** (c) as thin films deposited on ITO before and after thermal treatment. Solvent: 0.1 M TBAPF$_6$ in acetonitrile. Potential calculated versus ferrocene. Scan rate: 100 mV s^{-1}; T = 20 °C.

Tab. 3.8. Optical and electrochemical data of P-Boc-ThDPP2 - 4.[a]

	UV/nm	Onset of oxidation / V	Onset of reduction / V	HOMO{LUMO} / eV	HOMO-LUMO gap / eV
P-Boc-ThDPP2	550, 612	+0.50	-1.44	-5.30{-3.36}	1.95
P-Boc-ThDPP2 after TT	547, 600	+0.58	-1.26	-5.38{-3.54}	1.84
P-Boc-ThDPP3	560, 600	+1.01	-1.17	-5.81{-3.63}	2.18
P-Boc-ThDPP3 after TT	568, 595	+0.64	-1.28	-5.44{-3.52}	1.92
P-Boc-ThDPP4	552, 628	+0.58	-1.20	-5.38{-3.60}	1.78
P-Boc-ThDPP4 after TT	548, 602	+0.42	-1.00	-5.22{-3.80}	1.42

[a]Absorption spectra were taken in thin films. All redox potentials are referenced to the ferrocene/-ferrocenium redox couple. TT: thermal treatment.

In general, the polymers show lower HOMO- and higher LUMO-levels after the thermal treatment, which results in lower HOMO-LUMO gaps for the polymers without the Boc-substituents (Tab. 3.8). The origin for the lower gap is the same as for **P-Boc-ThDPP1**. After removal of the Boc-substituents, the strong hydrogen bonding is reconditioned leading to a more ordered structure and a stronger $\pi-\pi^*$-interaction between the polymer chains. The better extension of the conjugated polymer chain results in a lower energetic level.

In summary, a series of conjugated polymers based on Boc-substituted thienyl-DPP were synthesized via palladium-catalyzed Suzuki-coupling polycondensation reactions. It has been shown, that the removal of the Boc-substituent influences the optical and the electrochemical properties of the polymers. In general, after removal of the Boc-groups the polymers show higher HOMO- and lower LUMO-levels, resulting in lower HOMO-LUMO gaps. After thermal treatment, the polymer chains become more ordered due to a strong hydrogen bonding between the DPP units. It could be of great interest to use the cleavability of Boc-groups to optimize conjugated polymers for electronic applications.

3.5.3 New thienyl-DPP

New monomers based on 3-thienyl DPP were synthesized, and characterized using UV/vis-, and fluorescence-spectroscopy. The synthetic route to the monomers and polymers are described in Scheme 3.9.

Synthesis and characterization of the monomers

The monomer synthesis required a double cyclization of the 2 eq. thiophene-3-carbonitrile (**1**) and 1 eq. dimethyl succinate (**2**), yielding the pigment 3,6-di(thiophen-3-yl)pyrrolo[3,4-c]pyrrole-1,4(2H,5H)-dione (**3-ThDPP**) in 75%, which was alkylated with 3-(bromomethyl)heptanes, resulting the soluble monomer **EH-3-ThDPP** in the yield of 66%. The ^{1}H-NMR spectra displayed all the typical proton signals as expected. The peaks in the range from 7.8 to 8.7 ppm can be ascribed to the proton signals of the thiophene units, and the intensive signals between 0.7 and 1.6 ppm can be ascribed to the proton signals of the alkyl groups. Significantly, the duplet peak at 4.0 ppm can be ascribed to the α-CH$_2$ groups adjacently attached to the nitrogen atom of the lactam group in DPP (see Chapter 5, Experimental Part).

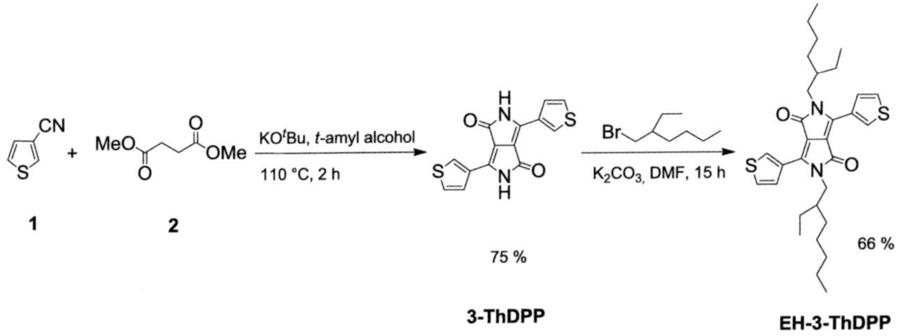

Scheme 3.9. Synthetic route to the monomers.

UV/vis absorption and fluorescence spectra of the monomers

The UV/vis absorption and fluorescence spectra of the monomers are shown in Fig. 3.38, and the optical data are listed in Tab. 3.9. The monomers show similar absorption bands with maxima in the range from 470 to 510 nm, and emission bands with maxima in the range from 510 to 550 nm. They show a greenish yellow colour in solution. From the absorption onsets at about 535 nm, optical HOMO-LUMO gaps of around 2.30 eV can be calculated, respectively. The photographs of the monomers in DCM solution are shown in Fig. 3.39.

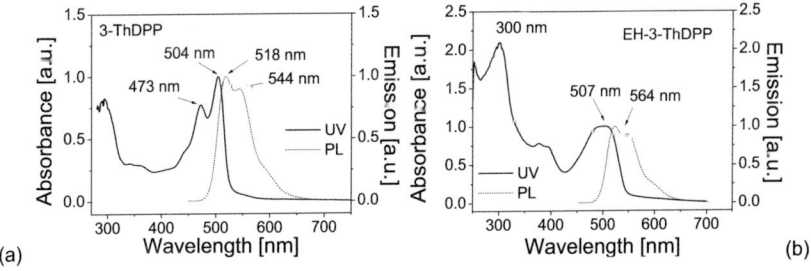

Fig. 3.38. UV/vis absorption and fluorescence spectra of **3-ThDPP** (a) in DMSO and **EH-3-ThDPP** (b) in DCM solutions.

Fig. 3.39. Photographs of **3-ThDPP** in DMSO and **EH-3-ThDPP** in DCM solutions

Compared to the 2-thienylDPPs, which exhibit UV/vis absorption maxima at about 530 nm, 3-thienyl-based DPP monomers are hypsochromically shifted. Interestingly, the 3-thienylDPP monomers show similar optical properties as the phenyl-based DPPs[15] with an absorption maximum at about 500 nm. The reason could be the angle between the 3-theinyl-rings and the DPP core. The molecule of 3-TheinylDPP is not as planar as the 2-thienylDPP, so that the conjugation system in 3-thienylDPP is not as extended as in 2-thienylDPP.

In summary, new 3-thienylDPP monomers were synthesized and characterized. In the molecule the thiophene unit is attached to the DPP core in the 3-position, which results in two reactive centres in the 2- and 5-positions of the thienyl-group to occur. Both positions can be functionalized. This could be of interest for further syntheses such as preparation of star-shaped macromolecules, or cross-linked polymers, which are suitable for optical or electronic applications.

3.6 Conclusion

A series of conjugated linear and cross-linked polymers based on 2,3,5,6-arylated diketopyrrolopyrrole (DPP) monomers was synthesized using anodic electrochemical oxidation and metal-mediated or -catalyzed polycondensation reactions such as Suzuki, Yamamoto and Sonogashira cross-coupling. It was shown that the incorporation of the tetra-functional DPP chromophore in a linear polymer chain via its 2- and 5- aryl units (i.e., the N-aryl units) results in non-π-conjugated polymers exhibiting HOMO-LUMO band gaps of about

3 eV, whereas incorporation via the aryl units in the 3- and 6-positions results in π-conjugated polymers with a bathochromic shift of the optical absorption by about 120 nm. The polymers show low HOMO-LUMO gaps in the range from 1.3 to 2.0 eV. The cross-linked polymers prepared using electropolymerization under anodic conditions show a very broad absorption with maxima around 510 nm, exhibiting low oxidation onset potentials in the range from -0.25 to -0.09 eV. They exhibited HOMO-LUMO band gaps of about 1.30 eV. They also show electrochromic properties under oxidation.

The polymer networks prepared by metal-mediated or -catalyzed cross-coupling polycondensation show microporous properties exhibiting BET surface areas up to 500 $m^2 g^{-1}$. The micoporosities of the networks are in the range from 17 to 38%.

In addition, thienylDPP monomers were substituted with *t*-butoxycarbonyl (Boc)-group and a series of conjugated polymers was synthesized via palladium-catalyzed Suzuki cross-coupling polycondensation reaction. A special emphasis was paid to the thermal cleavage of the Boc-group. The influence on the optical and electrochemical properties of the polymers was studied. It was shown that after removal of the Boc-groups the polymers generally exhibit higher HOMO- and lower LUMO-levels, resulting in lower HOMO-LUMO gaps. After thermal treatment, the polymer chains become more ordered due to a strong hydrogen bonding between the DPP units. It could be of great interest to use the cleavability of Boc-groups to adjust the electrochemical property and optimize conjugated polymers for electronic applications.

4 Conjugated polymers based on benzodifuranone

4.1 Symmetrical benzodifuranone-based conjugated polymers prepared via electropolymerization

The synthetic route to new benzodifuranone- (BZDF)-based monomers is described in Scheme 4.1. The starting compounds **1a-d** were prepared according to literature procedures (see Chapter 5, Experimental Part). The key compounds **BZDF1-4** were synthesized as shown in Scheme 4.1. In general, the synthesis of the symmetrical monomers **BZDF1-4** required the condensation of 0.5 eq. hydroquinone and 1 eq. of a mandelic acid derivative (**1a-d**), respectively, leading to a double cyclization of benzodihydrofurofurandione **2a-d**, followed by oxidation to the conjugated benzodifuranones **3a-d** in a yield of 65 to 75%. A Dean Stark apparatus was used for separation of water as byproduct favoring the double cyclization of dihydrofuranone in **2a-d**. Monomer **BZDF1** is a yellow solid and soluble in common organic solvents. In comparison, the bifunctionalized monomers **BZDF2-4** were greenish solids, they were less soluble in common solvents, and were not used for further palladium-catalyzed polymerization reactions such as Suzuki condensation at low temperatures. Only UV/Vis absorption and fluorescence spectra were recorded.

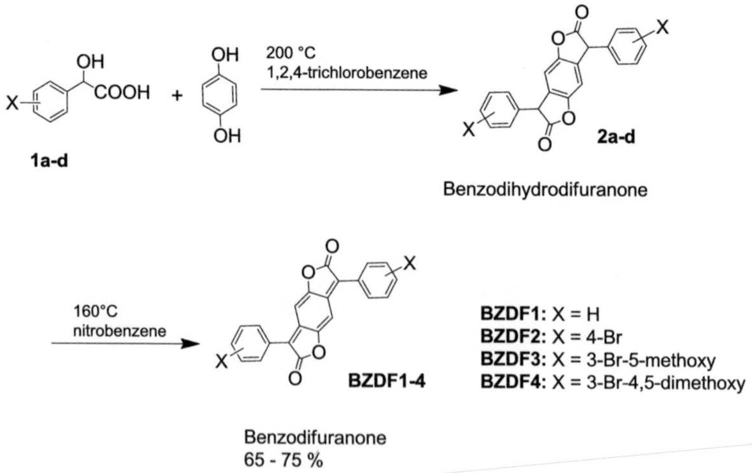

Scheme 4.1. Synthetic route to symmetrical BZDF monomers.

UV/Vis absorption and fluorescence spectra of monomers

The monomers **BZDF1-4** show strong absorption and weak emission in the visible with large Stokes shifts in a range from 80 to 140 nm. The optical data are listed in Table 3.1 and the UV/vis absorption and emission spectra are displayed in Fig. 4.1.

Table 3.1. Optical data of monomers **BZDF1 - 4** in DCM slolutions.

Monomers	Absorption λ_{max} / nm	Emission λ_{max} / nm	Optical HOMO-LUMO Gap / eV
BZDF1	469	600	2.36
BZDF2	438	586	2.30
BZDF3	391, 477	610	2.25
BZDF4	382, 521	606	2.10

From the absorption edges of the monomers in a range from 526 to 590 nm similar optical HOMO-LUMO gaps between 2.10 and 2.36 eV can be calculated. Monomer **BZDF4** with additional π-electrons of bromo- and methoxy-groups shows the smallest HOMO-LUMO gap of 2.10 eV. In comparison, monomer **BZDF1** with no substituent shows the largest HOMO-LUMO gap of 2.36 eV among the four monomers.

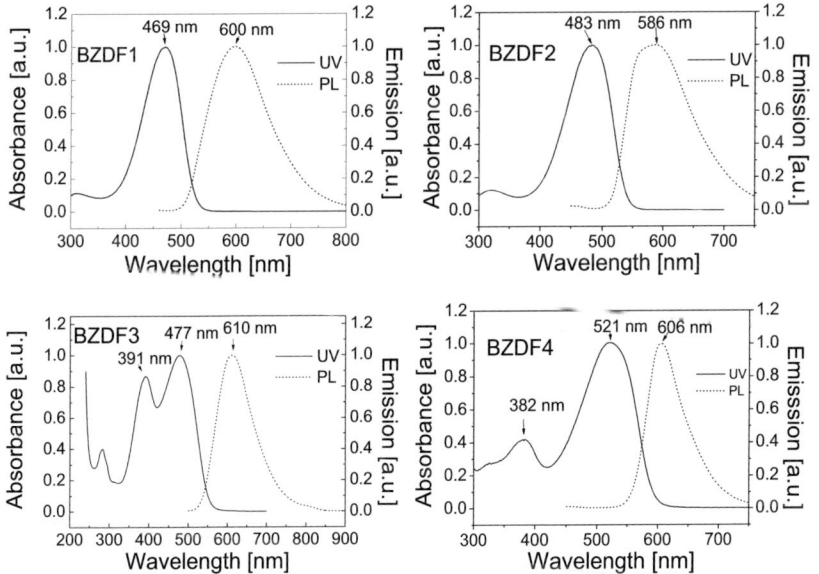

Fig. 4.1. UV/vis absorption and fluorescence spectra of monomers **BZDF 1-4**.

However, the low solubility of the bifunctionalized BZDF monomers can be overcome by heating the compounds in high boiling solvents such as DMF and DMSO at temperatures above 100 °C, so that the monomers are dissolved and can be used for synthetic procedures. This allowed us to prepare a number of BZDF monomers with aromatic solubility-increasing substituents as described in the following.

4.1.1 3,7-Diphenylbenzo[1,2-b:4,5-b']difuran-2,6-dione-based conjugated polymers prepared via electropolymerization

Electropolymerisation represents an alternative route to prepare such polymers in a catalyst-free approach. In this chapter, two electropolymerizable units, 3,4-ethylenedioxythiophene (EDOT) and 3,4-ethylenedithiathiophene (EDTT) were used as the solubility-increasing groups attached to the BZDF units.

4.1.1.1 Synthesis and properties of electropolymerizable monomers

The synthetic route to the electropolymerizable monomers is displayed in Scheme 4.2. **EDOT-BZDF** and **EDTT-BZDF** were synthesized via a microwave assisted Stille coupling of monomer **BZDF2** with 3,4-ethylenedioxythien-2-yl trimethylstannane (**EDOT1**) and 3,4-ethylenedithiathien-2-yl trimethylstannane (**EDTT1**) using $Pd(PPh_3)_4$ as the catalyst and DMF as the solvent. The coupling gave high yields of 79% for **EDOT-BZDF** and 89% for **EDTT-BZDF**. The two monomers are dark bluish solids being soluble in common solvents such as chloroform, dichloromethane, DMF, THF and toluene, for example.

The ^{1}H-NMR spectra of **EDOT-BZDF** and **EDTT-BZDF** displayed all the expected resonances with no discernible peaks corresponding to impurities (see Chapter 5, Experimental Part). Significantly, the singlet signal at about 7.0 ppm can be ascribed to the thiophene unit in the EDOT- and EDTT- substituents. The two triplet signals with a chemical shift of about 4.34 ppm are typical for the ethylene bridge of the EDOT unit. The two triplet signals at about 3.37 ppm are typical for the ethylene bridge of the EDTT unit, which are similar to those reported in the literature.[185]

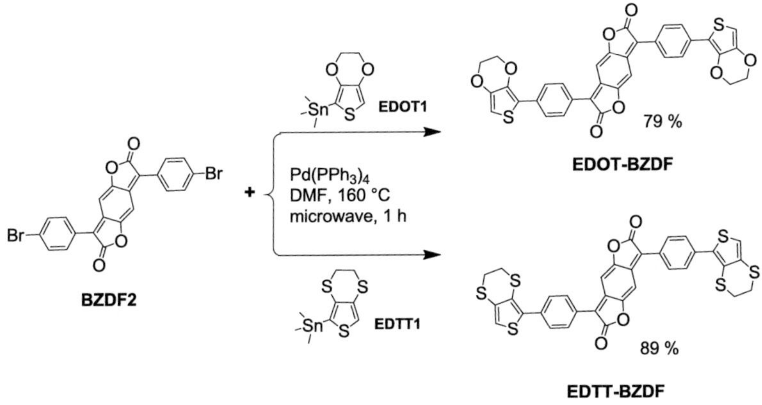

Scheme 4.2. Synthetic route to **EDOT-BZDF** and **EDTT-BZDF**.

UV/vis absorption of monomers

In the UV/vis absorption spectra of **EDOT-BZDF** and **EDTT-BZDF**, a large bathochromic shift is introduced via the addition of EDOT and EDTT to the conjugated system. Compared with the starting compounds **BZDF2**, the shifts are 154 nm for **EDOT-BZDF** and 64 nm for **EDTT-BZDF**. The spectra are displayed in Fig. 4.2. In dichloromethane **EDOT-BZDF** exhibits an absorption maximum at 590 nm, **EDTT-BZDF** at 502 nm (Tab. 4.2), respectively. Interestingly, the EDTT-substituted monomer **EDTT-BZDF** is not shifted bathochromically as largely as the EDOT-substituted monomers **EDOT-BZDF**. The reason could be a larger electron withdrawing effect caused by sulfur in the EDTT unit compared to oxygen in the EDOT units. As a consequence the conjugated system of the monomer **EDTT-BZDF** could become less enlarged. In comparison, EDOT shows a stronger electron donor effect, which could lead to a larger conjugation in **EDOT-BZDF**. These two monomers exhibit very strong absorption bands with an extinction coeffient of 63 580 L mol^{-1} cm^{-1} for **EDOT-BZDF** and 79 220 L mol^{-1} cm^{-1} for **EDTT-BZDF**. From the absorption edge of the monomers at about 690 nm optical HOMO-LUMO gaps between 1.80 and 1.90 eV can be calculated, respectively (Tab. 4.2). The photographs of **EDOT-BZDF** and **EDTT-BZDF** in dichloromethane solution are shown in Fig. 4.3.

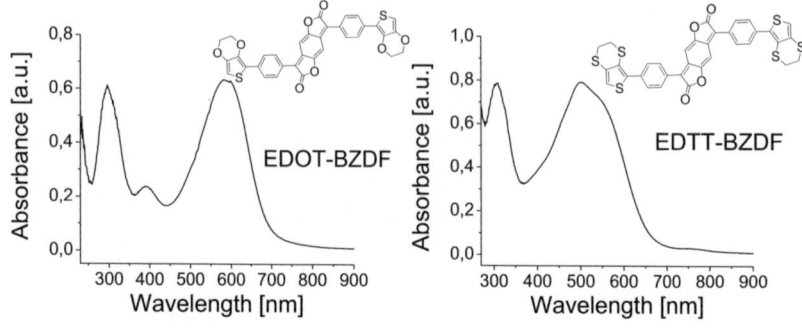

Fig. 4.2. UV/Vis spectra of **EDOT-BZDF** and **EDTT-BZDF**.

Tab. 4.2. Optical data of monomers **EDOT-BZDF** and **EDTT-BZDF** (in DCM).

	Absorption λ_{max} / nm	Optical HOMO-LUMO gap / eV	Extinction coefficient ε / L mol^{-1}cm^{-1}
EDOT-BZDF	298, 391, 590	1.81	63 580
EDTT-BZDF	306, 502	1.88	79 220

EDOT-BZDF   EDTT-BZDF

Fig. 4.3. Photographs of **EDOT-BZDF** and **EDTT-BZDF** in DCM solution.

Electrochemistry

Cyclic voltammograms of monomers

The electrochemistry of monomers **EDOT-BZDF** and **EDTT-BZDF** was studied using cyclic voltammetry in dichloromethane solution. The electrochemical details are given in Chapter 4, Experimental Part. The cyclic voltammograms of the monomers are shown in Fig. 4.4.

The electrochemical data are compiled in Tab. 4.3. For the monomer **EDOT-BZDF**, the oxidative cycle exhibits an irreversible peak at +0.91 V. The reductive cycle shows two reversible waves at -0.61 / -0.55 V and -0.94 / -0.88 V. **EDTT-BZDF** shows an irreversible oxidation peak at +0.84 V and two reversible cationic waves at -0.61 / -0.55 V and -0.93 / -0.87 V.

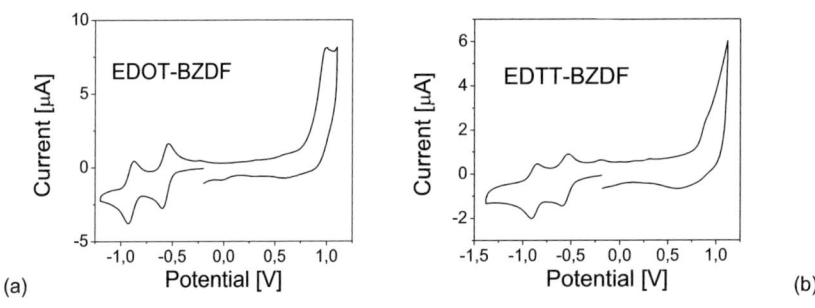

Fig. 4.4. Cyclic voltammograms of **EDOT-BZDF** (a) and **EDTT-BZDF** (b). Solvent: 0.1 M TBAPF$_6$/DCM. Potential calculated versus ferrocene. Scan rate: 100 mV s^{-1}; T = 20 °C.

Tab. 4.3. Electrochemical data of monomers **EDOT-BZDF** and **EDTT-BZDF**.[a]

	Onset of oxidation / V	HOMO / eV	Onset of reduction / V	LUMO / eV	HOMO-LUMO gap / eV
EDOT-BZDF	+0.83	-5.63	-0.50	-4.30	1.33
EDTT-BZDF	+0.82	-5.62	-0.47	-4.33	1.29

[a] All redox potentials are referenced to the ferrocene/ferrocenium redox couple. HOMO-LUMO gap according to the equation[177] -E_{LUMO} = $E_{onset(red)}$ + 4.8 eV and -E_{HOMO} = $E_{onset(ox)}$ + 4.8 eV, where $E_{onset(ox)}$ and $E_{onset(red)}$ are the onset potentials for the oxidation and reduction processes vs. ferrocene.

Interestingly, a high stability of the anion radical and the dianion showing reversible reductive waves was observed. This can possibly be explained with a stabilizing effect arising from the negative charges being at the oxygen atoms of the carbonyl groups of both furanone rings in the benzodifuranone monomers. This shows a very strong acceptor character of benzodifuranone. A suggested reduction mechanism is shown in Fig. 4.5. It also should be noted that the monomers showed small band gaps of about 1.30 eV, which are significantly low.

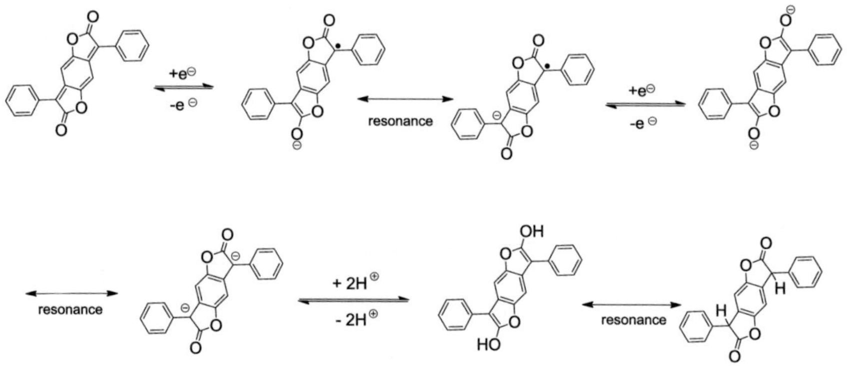

Fig. 4.5. A suggested reduction mechanism for benzodifuranone monomers.

4.1.1.2 Polymers prepared via electropolymerization

Electropolymerization

The two monomers **EDOT-BZDF** and **EDTT-BZDF** were electropolymerized using the same conditions as for the cyclic voltammetric experiments i.e., repetitive cycling over the redox active range of the materials (Scheme 4.3).

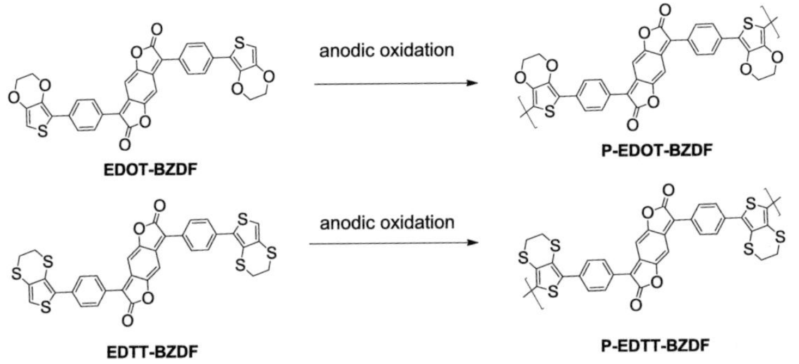

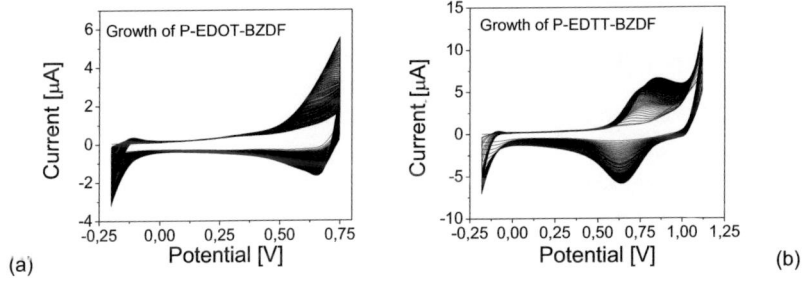

Scheme 4.3. Electropolymerization of **EDOT-BZDF** and **EDTT-BZDF**.

The polymer growth plots for the series are displayed in Fig. 4.6. **EDOT-BZDF** and **EDTT-BZDF** polymerized readily by continuous scanning between -0.20 and +1.10 V.

Fig. 4.5. Growth of P-**EDOT-BZDF** (a) and **P-EDTT-BZDF** (b) by cyclic voltammetry in dichloromethane using a carbon working electrode, and Ag wire pseudo-reference electrode. Supporting electrolyte: 0.1 M TBAPF$_6$. Scan rate: 100 mV s^{-1}; T = 20 °C.

New peaks appeared in lower voltage areas. This indicates a lower energy level of the resulting polymers, which should be caused by the elongation of the conjugated system in the polymer chains.

For the growth of both monomers, a plot of scan rate vs. current gives a linear fit (Fig. 4.67 confirming that charge transport through the film is not diffusion limited.

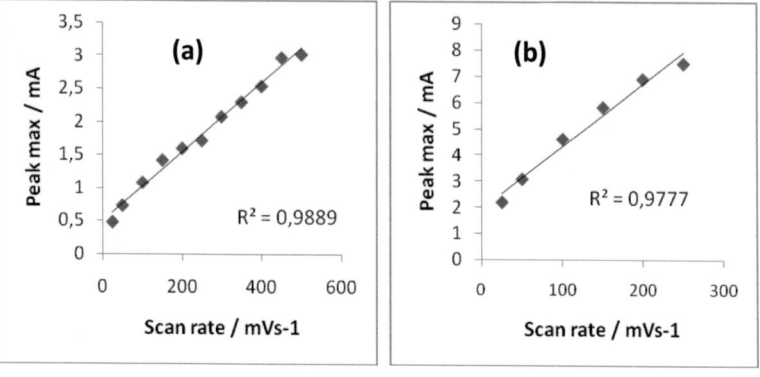

Fig. 4.7. Scan rate study of polymers **P-EDOT-BZDF** (a) and **P-EDTT-BZDF** (b).

Cyclic voltammograms of polymers

The cyclovoltammetric response of the polymers was studied using films of the polymers on a glassy carbon working electrode in acetonitrile vs. Ag/AgCl. The CV diagrams of **P-EDOT-BZDF** (a) and **P-EDTT-BZDF** (b) are shown in Fig. 4.8 and the electrochemical data are listed in Tab. 4.4. The two polymers show a quasi-reversible oxidative and two reversible reductive waves. The higher stability under reductive conditions was the same as for the monomers. The polymers also show low onset oxidation potentials and high onset reduction potentials, which lead to low HOMO- and high LUMO-levels for all the four polymers (see Tab. 4.4). This indicates notably small electrochemical band gaps for all four polymers with 0.44 eV for **P-EDOT-BZDF** and 0.91 eV for **P-EDTT-BZDF**, respectively. The very small band gaps indicate very strong donor-acceptor character of the polymers. The oxidative waves indicate the donor character of the thiophene derivatives EDOT and EDTT and the reduction waves the acceptor character of the benzodifuranone core.

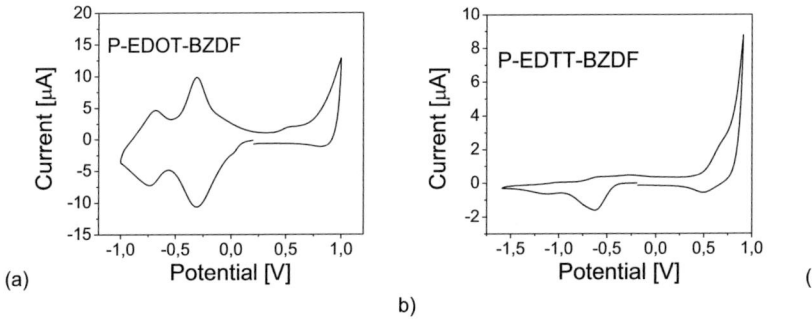

Fig. 4.8. Cyclic voltammograms of **P-EDOT-BZDF** (a) and **P-EDTT-BZDF** (b) as thin films electrodeposited on a glassy carbon electrode. Solvent: 0.1 M TBAPF$_6$/acetonitrile. Potential calculated versus ferrocene. Scan rate: 100 mV s^{-1}; T = 20 °C.

Spectroelectrochemistry

The absorption spectra of the polymers were taken from thin films on ITO-coated glass substrates. The spectra are displayed in Fig. 4.9 and the absorption data are listed in Tab. 4.4. The UV/vis absorption spectra of the polymers show very broad bands with no clear edges so that optical band gaps could not be calculated. The electrochemical band gap of the polymer is so small that the onset of the highest absorption edge would be outside the wavelength range of our device. Compared with the monomers, a large bathochromic shift occured for the polymers.

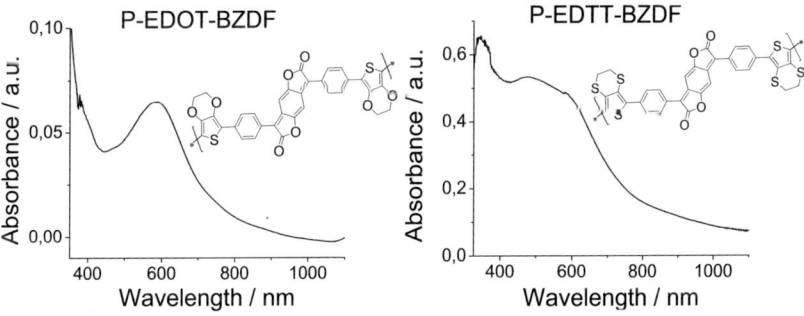

Fig. 4.9. UV/Vis spectra of polymers **P-EDOT-BZDF** and **P-EDTT-BZDF**.

Besides absorption maxima at 589 nm for **P-EDOT-BZDF** and 417 nm for **P-EDTT-BZDF**, the two polymers showed very broad shoulders ranging from about 680 to 750 nm. This indicates an elongation of the $\pi-\pi^*$-conjugated system in direction of the polymer chains.

Tab. 4.4. Optical and Electrochemical Data for Polymers **P-EDOT-BZDF** and **P-EDTT-BZDF**.[a]

	UV / nm	Onset of oxidation / V	Onset of reduction / V	HOMO{LUMO}+ / eV	HOMO-LUMO gap / eV
P-EDOT-BZDF	589	+0.41	0.03	-5.21{-4.77}	0.44
P-EDTT-BZDF	417, 600	+0.50	-0.41	-5.30{-4.39}	0.91

[a] Absorption spectra were taken from thin films.

Because of the high stability under reduction, it was of great interest to study the spectroelectrochemical properties for this series of polymers. For this study, the polymer films were grown on ITO-coated glass substrates and the electronic absorption spectra were taken with a gel electrolyte. 3-D absorption spectroelectrochemical plots for **P-EDOT-BZDF** and **P-EDTT-BZDF** are shown in Fig. 4.10. The absorption of the polymers under anodic oxidation did not show any changes, it was not studied in the following.

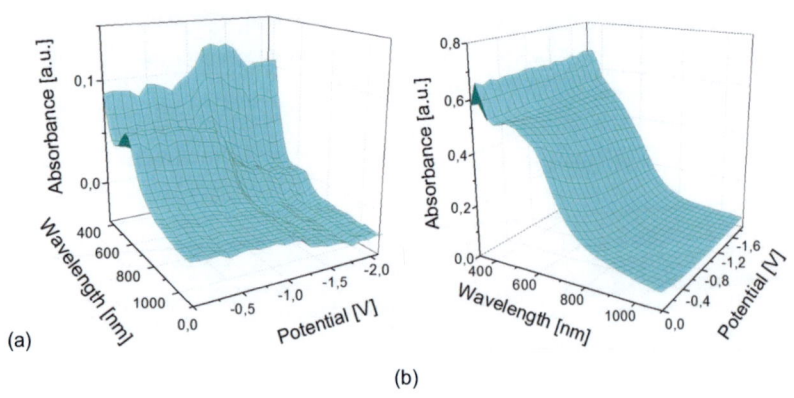

Fig. 4.10. 3-D absorption spectroelectrochemical plots for cathodic reduction of **P-EDOT-BZDF** (a) and **P-EDTT-BZDF** (b) as thin films on ITO in gel electrolyte. Ag wire pseudo-reference electrode. Potential calculated versus ferrocene. Scan rate: 100 mV s^{-1}; T = 20 °C.

The absorption of **P-EDOT-BZDF** did not change much until a potential of -1.0 V was reached. From -1.0 to -1.3 V a large drop of absorption emerged, which could point to the reduction from mono- to dianion involving a break in the conjugated system. Starting at -1.0 V, the film becomes more transparent. For **P-EDTT-BZDF**, it can be seen that all of the change in absorption spectra takes place during the first ten measurements until a potential of -0.7 V is reached. This is just past the first reduction peak in the CV. By taking it to -1.9 V, there is only little more change in the absorption which corresponds to the fact that the second reduction peak is smaller compared to the first one. The colour change was only small, the film turned from a dark to a lighter purple/brown colour. The colour of the film changed from dark purple to light orange, transparent colour. For both polymers, the 2-D representation of electronic absorption spectra gives a better view of decreasing absorption maxima under reduction conditions (Fig. 4.11).

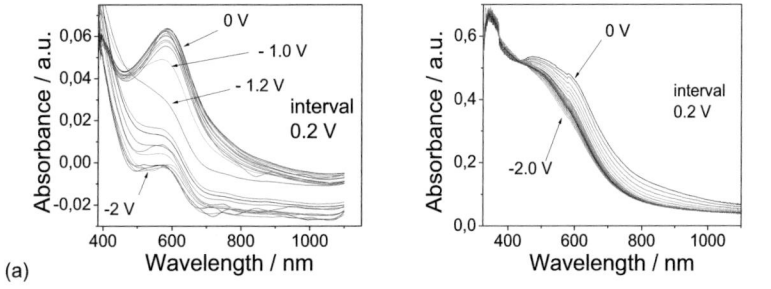

Fig. 4.11. Absorption spectroelectrochemical plots for **P-EDOT-BZDF** (a) and **P-EDTT-BZDF** (b). Potential decreases with an interval of 0.2 V.

In summary, the synthetic route and the key properties of a series of electropolymerizable monomers containing the benzodifuranone moiety was described. We also have synthesized the first polymers based on EDOT- or EDTT- benzodifuranone units electrochemically. The polymers exhibit a broad absorption, high HOMO- and low LUMO-levels and very small band gaps. Interesting electrochromic properties were found under oxidative and reductive conditions. Under reduction, the polymers showed colour changes from dark to almost transparent. Due to the broad absorption bands, the very small band gaps, and the reversibility of oxidation and reduction processes, **P-EDOT-BZDF** and **P-EDTT-BZDF** might be useful for electronic applications.

4.1.2 3,6-Diphenylbenzo- [1,2-b:6,5-b']difuran-2,7-dione-based conjugated polymers prepared via electropolymerization

An isomer of 3,7-diphenylbenzo[1,2-b:4,5-b']difuran-2,6-dione (*para*-form) can be synthesized, if 0.5 eq. 1,2-dihydroxybenzene are used for the condensation with 1 eq. of a mandelic acid derivative instead of using hydroquinone. The synthetic route to the *ortho*-isomer, 3,6-diphenylbenzo[1,2-b:6,5-b']difuran-2,7-dione (*ortho*-form), is displayed in Scheme 4.4.

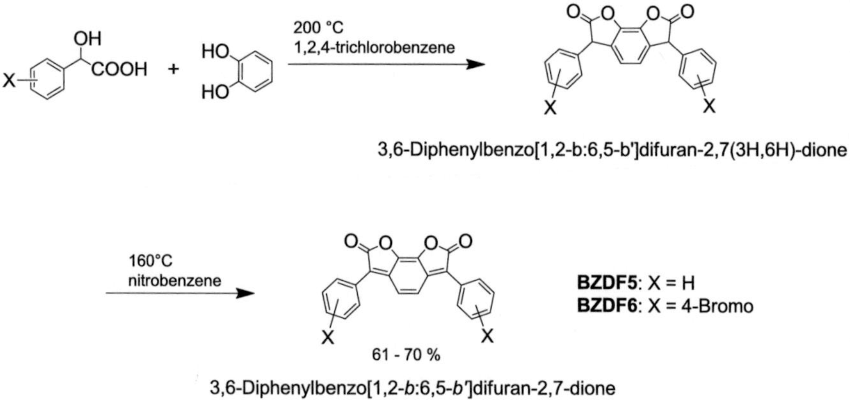

3,6-Diphenylbenzo[1,2-b:6,5-b']difuran-2,7(3H,6H)-dione

BZDF5: X = H
BZDF6: X = 4-Bromo

3,6-Diphenylbenzo[1,2-*b*:6,5-*b'*]difuran-2,7-dione

Scheme 4.4. Synthetic route to the *ortho*-formed monomers **BZDF5** and **6**.

Both monomers were red solids. Compared to the unsubstituted monomer **BZDF5**, the bifunctionalized monomer **BZDF6** was less soluble in common organic solvents. However, **BZDF6** showed a better solubility compared to the *para*-formed monomer **BZDF2**, this could be caused by its angled structure, which is not advantageous fo the formation of a close-packed structure in the solid state. Both monomers showed strong absorptions with maxima at about 462 nm for **BZDF5** and 492 nm for **BZDF6**. The UV/vis absorption spectra are shown in Fig. 4.12, and the optical data are listed in Tab. 4.5.

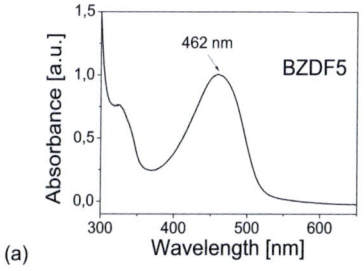 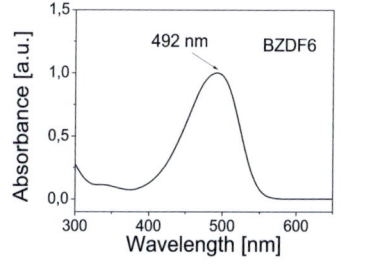

Fig. 4.12. UV/Vis spectra of compounds **BZDF5** (a) and **BZDF6** (b).

From the absorption edges at about 518 nm for **BZDF5**, and 547 nm for **BZDF6**, optical HOMO-LUMO gaps of 2.40 eV for **BZDF5**, and 2.27 eV for **BZDF6**, respectively, can be calculated. Photographs of both monomers in solution are shown in Fig. 4.13.

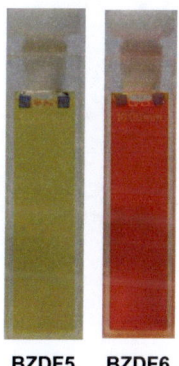

BZDF5 **BZDF6**

Fig. 4.13. Photographs of **BZDF5** and **6** in DCM solution.

Table 3.5. Optical data of monomers **BZDF5** and **6** in DCM.

Monomers	Absorption λ_{max} / nm	Optical HOMO-LUMO gap / eV
BZDF5	462	2.40
BZDF6	492	2.27

4.1.2.1 Synthesis and properties of electropolymerizable monomers

As an analogue to the benzo[1,2-b:4,5-b']difuran-2,6-dione-based monomers **EDOT-BZDF** and **EDTT-BZDF**, two new electropolymerizable benzo[1,2-b:6,5-b']difuran-2,7-dione-based monomers **EDOT-BZDF** and **EDTT-oBZDF** were synthesized via a Stille coupling of monomer **BZDF6** with **EDOT1** and **EDTT1** under the same reaction conditions (Scheme 4.5).

The microwave assisted Stille coupling gave high yields of 82% for **EDOT-oBZDF** and 86% for **EDTT-oBZDF**. These two monomers are dark bluish solid being soluble in common solvents such as chloroform, dichloromethane, DMF, THF and toluene, which is similar to the *para*-formed analogues **EDOT-BZDF** and **EDTT-BZDF**.

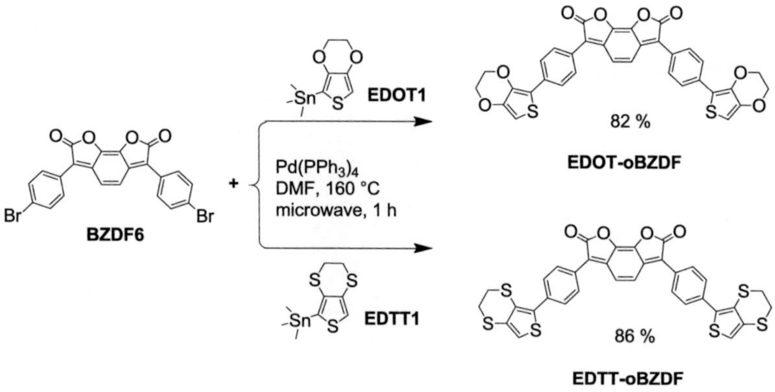

Scheme 4.5. Synthetic route to **EDOT-oBZDF** and **EDTT-oBZDF**.

Similar to the *para*-isomes, the ^{1}H-NMR spectra of *ortho*-isomers **EDOT-oBZDF** and **EDTT-oBZDF** displayed identical resonances with no discernible peaks corresponding to impurities (see Chapter 5, Experimental Part). Specifically, the singlet signal at about 7.0 ppm can be described to the thiophene unit in the EDOT- and EDTT- substituents. The two triplet signals with a chemical shift of about 4.34 ppm are typical for the ethylene bridge of the EDOT unit. The two triplet signals at about 3.27 ppm are typical for the ethylene bridge of the EDTT unit.

UV/vis absorption of monomers

The UV/vis absorption spectra of **EDOT-oBZDF** and **EDTT-oBZDF** indicate that a large bathochromic shift is introduced by the addition of EDOT and EDTT to the conjugated system. Compared with the starting compound **BZDF6**, the shifts are 95 nm for **EDOT-oBZDF**, and 65 nm for **EDTT-oBZDF**. The spectra are displayed in Fig. 4.14. In dichloromethane, **EDOT-OBZDF** exhibits an absorption maximum at 588 nm, and **EDTT-oBZDF** at 558 nm (Tab. 4.6), respectively. The EDTT- substituted monomer **EDTT-oBZDF** is not shifted bathochromically as largely as the EDOT- substituted monomer **EDOT-oBZDF**, which is similar to the *para*-isomeric monomers **EDOT-BZDF** and **EDTT-BZDF** (see Chapter 3.1.1.1). The reason could be a larger electron withdrawing effect caused by sulfur in the EDTT unit compared to oxygen in the EDOT units. This could lead to a lower extension of the conjugated system of the monomer **EDTT-oBZDF**. In comparison, EDOT shows a stronger electron donor effect, which could lead to a larger conjugation in **EDOT-oBZDF**.

From the nearly same absorption edge of the monomers at about 690 nm, similar optical HOMO-LUMO gaps between 1.80 and 1.90 eV (Tab. 4.6), respectively, can be calculated.

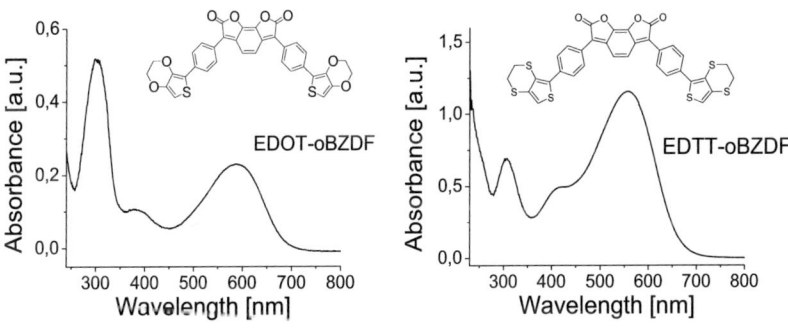

Fig. 4.14. UV/Vis spectra of **EDOT-oBZDF** and **EDTT-oBZDF**.

Significantly, these two monomers also exhibit very a strong absorption with extinction coeffients of 23 080 L mol^{-1} cm^{-1} for **EDOT-oBZDF**, and 115 560 L mol^{-1} cm^{-1} for **EDTT-oBZDF**. The photographs of the very dark solution of the two monomers are shown in Fig. 4.15. From the absorption edge of the monomers at about 690 nm optical HOMO-LUMO gaps between 1.80 and 1.86 eV, respectively, can be calculated (Tab. 4.6).

EDOT-oBZDF   EDTT-oBZDF

Fig. 4.15. Photographs of **EDOT-oBZDF** and **EDTT-oBZDF** in DCM solution.

Tab. 4.6. Optical data of monomers **EDOT-oBZDF** and **EDTT-oBZDF** (in DCM).

	Absorption λ_{max} / nm	Optical HOMO-LUMO gap / eV	Extinction coefficient ε / L mol^{-1}cm^{-1}
EDOT-oBZDF	302, 588	1.80	23 080
EDTT-oBZDF	307, 407, 558	1.86	115 560

Electrochemistry

Cyclic voltammograms of monomers

The electrochemistry of the monomers **EDOT-oBZDF** and **EDTT-oBZDF** was studied using cyclic voltammetry in dichloromethane solution. The cyclic voltammograms of the monomers are shown in Fig. 4.16. The electrochemical data are compiled in Tab. 4.7. **EDOT-oBZDF** shows an irreversible oxidation peak at +0.65 V and two reversible reductive waves at -0.93 / -0.86 V and -1.22 / -1.18 V. **EDTT-oBZDF** undergoes an irreversible oxidation at +1.00 V and the reduction shows two reversible waves at -0.55 / -0.49 V and -0.87 / -0.82 V.

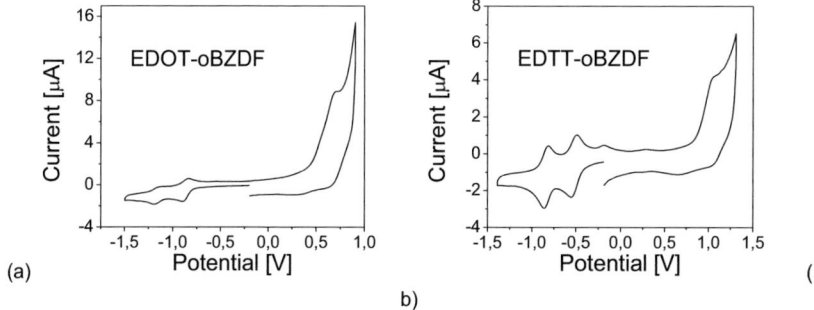

Fig. 4.16. Cyclic voltammograms of **EDOT-oBZDF** (a) and **EDTT-oBZDF** (b). Solvent: 0.1 M TBAPF$_6$ in DCM. Potential calculated versus ferrocene. Scan rate: 100 mV s^{-1}; T = 20 °C.

Similar to the *para*-isomer, a high stability of the anion radical of the two monomers **EDOT-oBZDF** and **EDTT-oBZDF** showing reversible reductive waves was observed. The same stabilizing effect arising from the negative charges being at the oxygen atoms in the carbonyl groups of both furanone rings in the benzodifuranone monomers can also emerge in the two monomers. This shows a very strong acceptor character of the *ortho*-isomeric benzodifuranones. Small band gaps of about 1.25 eV were calculated.

Tab. 4.7. Electrochemical data of monomers **EDOT-oBZDF** and **EDTT-oBZDF**. [a]

	Onset of oxidation / V	HOMO / eV	Onset of reduction / V	LUMO / eV	HOMO-LUMO gap / eV
EDOT-oBZDF	+0.45	-5.25	-0.80	-4.00	1.25
EDTT-oBZDF	+0.85	-5.65	-0.41	-4.39	1.26

[a] All redox potentials are referenced to the ferrocene/ferrocenium redox couple. HOMO-LUMO gap according to the equation[177] -E_{LUMO} = $E_{onset(red)}$ + 4.8 eV and -E_{HOMO} = $E_{onset(ox)}$ + 4.8 eV, where $E_{onset(ox)}$ and $E_{onset(red)}$ are the onset potentials for the oxidation and reduction processes vs. ferrocene.

4.1.2.2 Polymers prepared via electropolymerization

Electropolymerization

The two monomers **EDOT-oBZDF** and **EDTT-oBZDF** were electropolymerized by repetitive cycling over the redox active range of the materials using the same conditions as for **EDOT-BZDF** and **EDTT-BZDF** (Scheme 4.6).

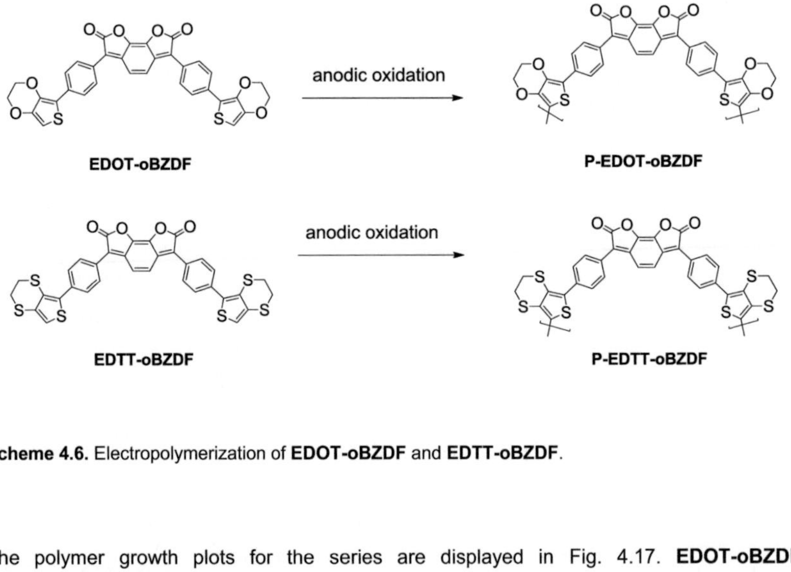

Scheme 4.6. Electropolymerization of **EDOT-oBZDF** and **EDTT-oBZDF**.

The polymer growth plots for the series are displayed in Fig. 4.17. **EDOT-oBZDF** polymerized readily, **EDTT-oBZDF** required more cycles. 120 cycles of continuous scanning between -0.20 and +1.10 V were needed for **EDTT-oBZDF**, before a satisfactory film was obtained, whereas 60 cycles were needed for **EDOT-oBZDF** in order to get a film of comparable quality. It should be noted that the electropolymerization of thiophenes, especially of EDTTs at relatively low oxidation potentials can lead to oligomeric products rather than to long chain polymers.[186] This could be the reason for the slow growth rate for **EDTT-oBZDF**. In comparison with **EDOT-oBZDF**, the EDTT-substituted **EDTT-BZDF** (see Chapter 3.1.1.1) exhibited a much better growth. In this case the angled structure of **EDTT-oBZDF** could be sterically unfavorable for a polymer growth in comparison with the rod-like structure of **EDTT-BZDF**.

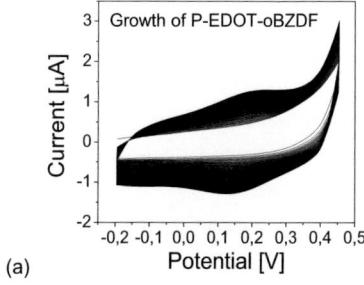
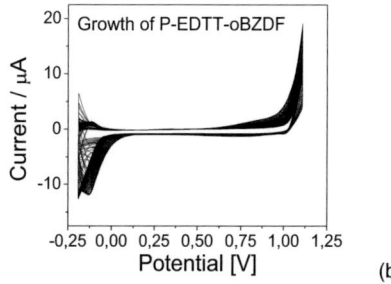

Fig. 4.17. Growth of **P-EDOT-oBZDF** (a) and **P-EDTT-oBZDF** (b) by cyclic voltammetry in dichloromethane using a carbon working electrode, and a Ag wire pseudo-reference electrode. Supporting electrolyte: 0.1 M TBAPF$_6$. Scan rate: 100 mV s^{-1}; T = 20 °C.

For all compounds, a plot of scan rate vs. current gives a linear fit confirming that charge transport through the film is not diffusion limited (Fig. 4.18).

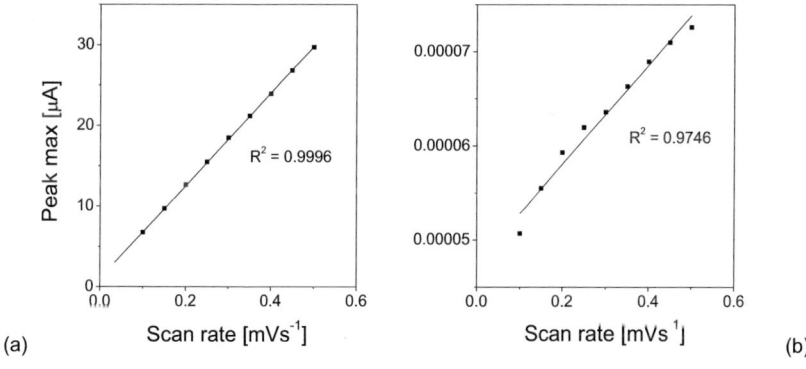

Fig. 4.18. Scan rate study of polymers **P-EDOT-oBZDF** (a) and **P-EDTT-oBZDF** (b).

Cyclic voltammograms of polymers

Under the same conditions as for the films of the *para*-isomers, the cyclovoltammetric response of the polymer films of the *ortho*-isomers was studied. Films of the polymers were deposited on a glassy carbon working electrode in acetonitrile and cycled using an Ag/AgCl pseudo-reference electrode. The CV diagrams of **P-EDOT-oBZDF** (a) and **P-EDTT-oBZDF** (b) are shown in Fig. 4.19 and the electrochemical data are listed in Tab. 4.8. **P-EDOT-oBZDF** shows a quasi-reversible oxidative and a reversible reductive wave. **P-EDTT-oBZDF** shows a irreversible oxidative and two reversible reductive waves. The only appearing reductive wave of **P-EDOT-oBZDF** at -0.65 V could be caused by the experiment voltage starting at -0.25 V. Compared to the *para*-isomer **P-EDOT-BZDF**, which exhibits the first reductive wave at -0.25 V, the first reductive wave of **P-EDOT-oBZDF** could also emerge at a high voltage similar to **P-EDOT-BZDF**. It could also be possible, that the first and the second reductive waves of **P-EDOT-oBZDF** are so near to each other, that they are molten into a single wave in the CV diagram.

The higher stability under reductive conditions was the same as for the monomers. The polymers also show low onset oxidation potentials and high onset reduction potentials, which lead to low HOMO- and high LUMO-levels for all polymers (see Tab. 4.8). This indicates notably small electrochemical band gaps for all four polymers (0.36 eV for **P-EDOT-oBZDF**, and 1.28 eV for **P-EDTT-oBZDF**), respectively. The polymers containing EDOT-substituted benzodifuranone show smaller electrochemical band gaps, lower HOMO- and higher LUMO-levels than the polymers containing EDTT-substituted benzodifuranone. This should be caused by the same reason as for the *para*-isomers, **P-EDOT-BZDF** and **P-EDTT-BZDF**. The polymer main chain containing EDOT is more planar than the polymer chain containing EDTT due to the oxygen-sulfur interaction between the adjacent EDOT units, as described in Chapter 3. EDTT does not provide an advantage for planarity (Fig. 4.20).

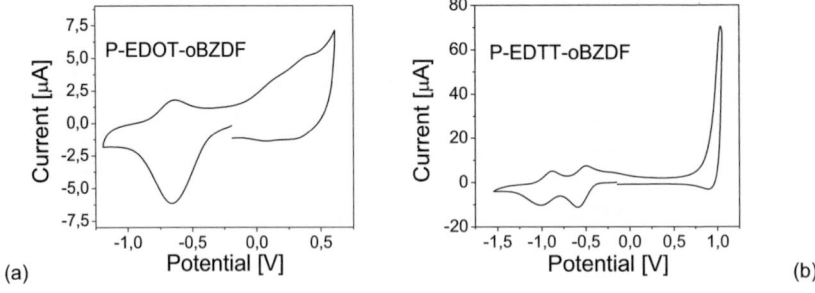

Fig. 4.19. Cyclic voltammograms of **P-EDOT-oBZDF** (a) and **P-EDTT-oBZDF** (b) as thin films electrodeposited on a glassy carbon electrode. Solvent: 0.1 M TBAPF$_6$/acetonitrile. Potential calculated versus ferrocene. Scan rate: 100 mV s^{-1}; T = 20 °C.

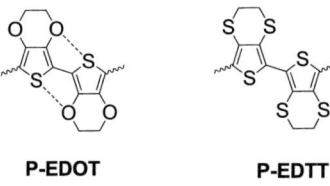

P-EDOT **P-EDTT**

Fig. 4.20. Interaction between the adjacent EDOT units, no interaction between the adjacent EDTT units.

The sterical difference between the angled and the rod-like structures could be another factor affecting the electronic properties. Compared with the monomer **EDTT-oBZDF**, the electrochemical band gap of **P-EDTT-oBZDF** with 1.28 eV show a similar band gap as for the monomer **EDTT-BZDF** with 1.26 eV. The reason could be that the polymer chain of **P-EDTT-BZDF** is not fully conjugated due to the angled structure of the monomer units. As a main reason, the lower planarity of **P-EDTT-oBZDF** could lead to a twisted polymer main chain. The polymer can be ascribed to a gathering of separate monomer units, which rather show similar optical and electrochemical properties as for the monomer **EDTT-oBZDF**.

Spectroelectrochemistry

The absorption spectra of the polymers were taken from thin films on ITO-coated glass substrates. The spectra are displayed in Fig. 4.21, and the absorption data are listed in Table 3.8. The UV/vis absorption spectra of the polymers show very broad bands with no clear edges so that optical band gaps could not be calculated.

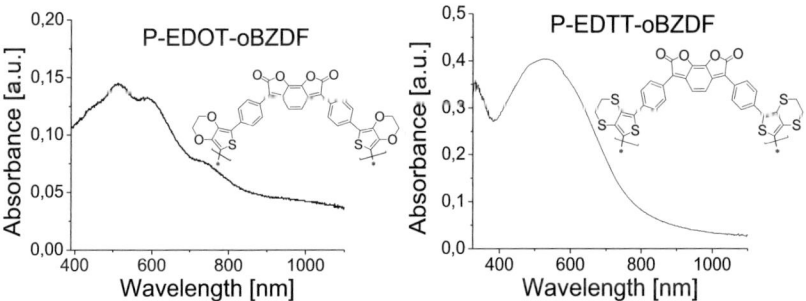

Fig. 4.21. UV/Vis spectra of polymers **P-EDOT-oBZDF** and **P-EDTT-oBZDF**.

The electrochemical band gap of the polymer is so small that the onset of the highest absorption edge is outside the wavelength range of our spectrometer. Compared with the monomers, a large bathochromic shift occured for the polymers. Besides absorption maxima at 587 nm for **P-EDOT-oBZDF**, and 530 nm for **P-EDTT-oBZDF**, the polymers showed very broad shoulders ranging from about 680 to 750 nm. This indicates an elongation of the π–conjugated system in direction of the polymer chains.

Tab. 4.8. Optical and Electrochemical Data for Polymers **P-EDOT-oBZDF** and **P-EDTT-oBZDF**.[a]

	UV / nm	Onset of oxidation / V	Onset of reduction / V	HOMO{LUMO} / eV	HOMO-LUMO gap / eV
P-EDOT-oBZDF	515, 587, 743	-0.01	-0.37	-4.79{-4.43}	0.36
P-EDTT-oBZDF	530, 680	+0.87	-0.41	-5.67{-4.39}	1.28

[a] Absorption spectra were taken from thin films (thickness ca. 100 nm).

It was of great interest to study the spectroelectrochemical properties for the polymers due to the high stability under reduction. Polymer films were grown on ITO-coated glass substrates and the electronic absorption spectra were taken with a gel electrolyte. A more detailed description of the experiment is given in Chapter 5, Experiment Part. 3-D absorption spectroelectrochemical plots for **P-EDOT-oBZDF** and **P-EDTT-oBZDF** are shown in Fig. 4.22. Similar to the *para*-isomers, the absorption spectra of **P-EDOT-OBZDF** and **P-EDTT-oBZDF** barely showed any changed under anodic oxidation. Only the absorption under cathodic reduction for the polymers was studied in the following. The 3D-plot of **P-EDOT-oBZDF** indicates a drop in the absorption band in the range from 514 to 593 nm at -0.5 V as the polymer goes from a purple colour to more transparent. The spectroelectrochemistry of **P-EDTT-oBZDF** indicates a drop in the $\pi-\pi^*$ transition at 520 nm at the first reductive wave at -0.8 V, and a new band at 770 nm forms, which is exptended into the near-IR. It might correspond to the formation of a radical anion in the polymer chain. At the second reductive wave at -1.0 V, the new band disappears again, probably due to formation of the dianion.

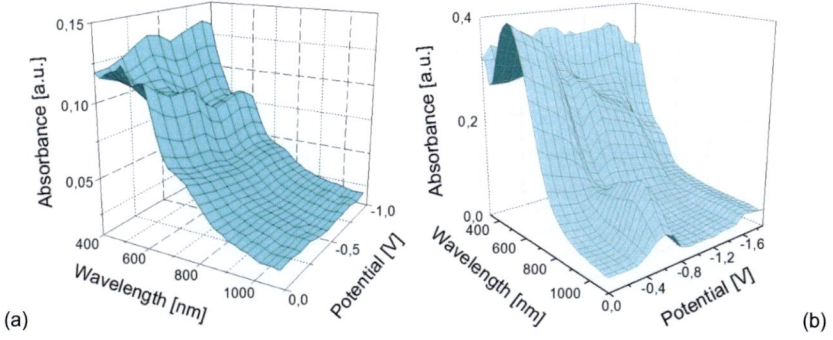

Fig. 4.22. 3-D absorption spectroelectrochemical plots for **P-EDOT-oBZDF** (a) and **P-EDTT-oBZDF** (b) as thin films on ITO in gel electrolyte. Ag wire pseudo-reference electrode. Potential calculated versus ferrocene. Scan rate: 100 mV s^{-1}; T = 20 °C.

Simultaneously, the absorption band at 520 nm decreases intensively, if the potential is gradually shifted towards -1.9 V. The colour of the film changes from dark purple to a light orange, transparent colour. The 2-D electronic absorption spectra shown in Fig. 4.23 give a direct view of the change of absorption maxima.

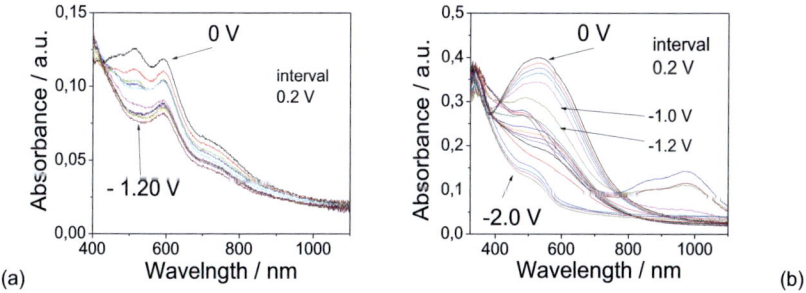

Fig. 4.23. 2-D absorption spectroelectrochemical plots for **P-EDOT-oBZDF** (a) and **P-EDTT-oBZDF** (b), potential decreases with an interval of 0.2 V.

Within all the four *para-* and *ortho-*isomeric polymers, the polymers **P-EDOT-BZDF** and **P-EDOT-oBZDF** containing EDOT moieties show lower HOMO-LOMO gaps that the EDTT substituted polymers **P-EDTT-BZDF** and **P-EDTT-oBZDF**. The reason is the higher planarity in both polymer main chains containing EDOT units, which leads to a better an extension of the $\pi-\pi^*$-conjugated system in direction of the polymer chains. Compared to *para-*isomeric polymers, the angled structure of the *ortho-*isomers **P-EDOT-oBZDF** and **P-EDTT-oBZDF** are neither advantageous for the polymer growth nor for the elongation of the $\pi-\pi^*$-conjugation.

In summary, the synthetic route and the key properties of a series of electropolymerizable monomers containing *ortho-*benzodifuranones was presented. The first polymers based on EDOT- or EDTT-, and *ortho-*benzodifuranone units were synthesized electrochemically. The polymers exhibit a broad absorption, high HOMO- and low LUMO-levels and very small band gaps (about 0.40 eV for **P-EDOT-oBZDF**). Interesting electrochromic properties were found under oxidative and reductive conditions. Under reduction, the polymers showed colour changes from dark to almost transparent. The polymers might be useful for electronic applications.

4.2 Symmetrical benzodifuranone-based conjugated polymers prepared via palladium-catalyzed cross-coupling polycondensation

For a regular process of the Suzuki cross-coupling polycondensation properly adjusted reaction conditions are required. The purity of the starting materials contributes strongly to the success of the coupling. Furthermore, a good solubility of monomers is of crucial importance to achieve high molecular weights. Therefore, the little soluble bifunctionalized monomers **BZDF2** and its *ortho-*formed isomer **BZDF6** are not suitable for a direct proceeding of both Suzuki and Stille polycondensation reactions.

In this chapter, bifunctionalized well-soluble benzodifuofuranone monomers based on **BZDF2** and **BZDF6** were synthesized, and new conjugated polymers containing benzofurofurandiketone (BZDF) units were prepared via palladium catalyzed Suzuki and Stille cross-coupling polycondensation reactions.

4.2.1 Conjugated polymers prepared via Suzuki cross-coupling polycondensation

4.2.1.1 Synthesis and properties of monomers

The synthetic route to the two well-soluble monomers is displayed in Scheme 4.7. **BZDF7** and **BZDF8** were synthesized via microwave assisted Stille coupling of monomers **BZDF2** and **BZDF6** with trimethyl(3-octylthiophen-2-yl)stannane (**OTH1**) using Pd(PPh$_3$)$_4$ as the catalyst and DMF as the solvent. The poor solubility of the starting compounds was overcome using a high reaction temperature of 160°C. The coupling proceeded in high yields of 90% for **BZDF7** and 84% for **BZDF8**, respectively. Both compounds were brominated using NBS, yielding the bifunctional monomers **BZDF9** and **BZDF10** in 80 and 75% yield, respectively. The 3-octyl chain of the thiophene ring was used as the solubility increasing group.

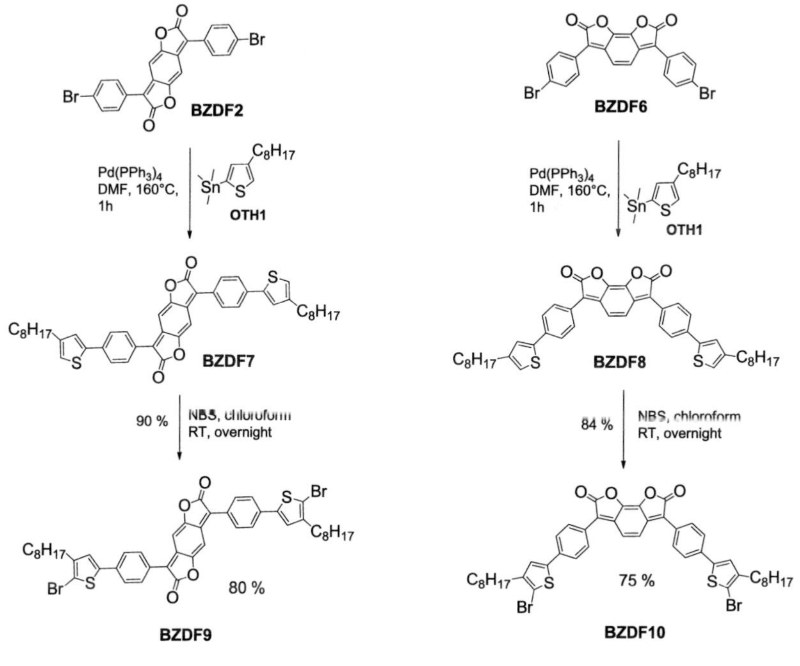

Scheme 4.7. Synthetic route to **BZDF9** and **BZDF10**.

The ¹H-NMR spectra of the four monomers displayed all the expected resonances with no discernible peaks corresponding to impurities (see Chapter 5, Experimental Part). The singulet signal at about 6.90 ppm can be ascribed to the protons in the 2-position of thiophene in the monomers **BZDF7** and **9**, which disappeared after bromation of the monomers **BZDF8** and **10** with NBS. The signals with chemical shifts between 0.80 and 2.70 ppm are typical for the 3-octyl groups. The triplet signals at about 2.67 ppm can be ascribed to the α-CH_2-protons of the 3-octyl group.

All the four monomers are dark solids, which are very soluble in common solvents such as chloroform, dichloromethane, DMF, THF and toluene, for example. The solutions are dark purple (Fig. 4.24).

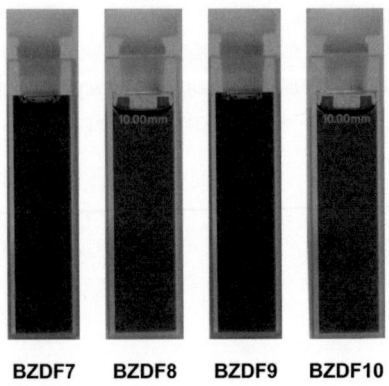

BZDF7   BZDF8   BZDF9   BZDF10

Fig. 4.24. Photographs of the solution of **BZDF 7-10** in DCM solution.

UV/vis absorption of the monomers

Compared with the starting compounds **BZDF2** and **6**, the absorptions of the four thiophene-substituted monomers **BZDF7 - 10** are strongly bathochromically shifted. The shifts are about 90 nm. This is caused by the addition of the thiophene units to the conjugated system. The spectra are displayed in Fig. 4.25, and the optical data are listed in Tab. 4.9. The four monomers **BZDF7 - 10** exhibit absorption maxima at about 560 nm. All the four monomers show high extinction coefficients in a range of 50 000 to 60 000 L mol^{-1} cm^{-1}. From the absorption onset at about 650 nm optical HOMO-LUMO gaps of about 1.90 eV can be calculated (Tab. 4.9).

Tab. 4.9. Optical data of monomers **BZDF7 - 10** (in DCM solution).

Monomers	Absorption λ_{max} / nm	Optical HOMO-LUMO gap / eV	Extinction coefficient ε / L mol^{-1} cm^{-1}
BZDF7	550	1.92	56 300
BZDF8	551	1.86	50 060
BZDF9	556	1.94	59 450
BZDF10	553	1.89	51 670

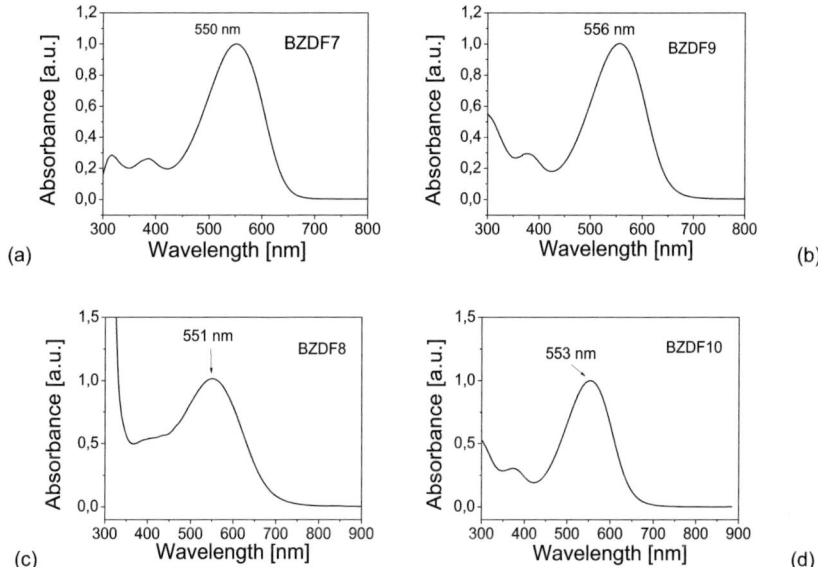

Fig. 4.25 UV/Vis absorption and emission spectra of **BZDF7 - 10** in DCM solution.

4.2.1.2 Synthesis and properties of the polymers

Using palladium-catalyzed Suzuki coupling, monomers **BZDF9** and **10** were polymerized with the comonomer 2,5-bis(4,4,5,5-tetramethyl-1,3,2-dioxaborolan-2-yl)thiophene (**TH1**) (Scheme 3.8).

The polymers are dark solids. They are soluble in common organic solvents such as dichloromethane, chloroform, toluene, or dimethyl formaldehyde, but less soluble in methanol. The ^{1}H-NMR spectra of two polymers show broad signals of the protons of the phenyl rings at 7.60 - 7.80 ppm, the signal of the protons of the two thiophene units at about 7.40 ppm, and signals of the alkyl protons in the range from 0.80 to 2.40 ppm. Most signals resemble the chemical shifts of the monomers **BZDF9** and **10**.

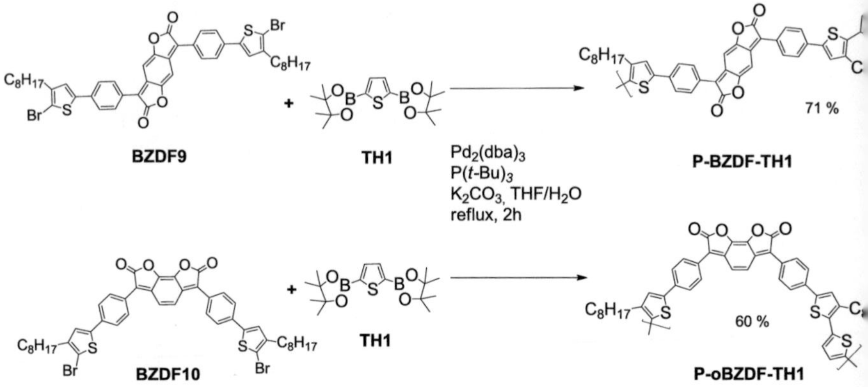

Scheme 4.8. Synthetic route to polymers **P-BZDF-TH1** and **P-oBZDF-TH**.

UV/vis absorption of the polymers

The UV/vis absorption spectra of both polymers are shown in Fig. 4.26, and the optical data are listed in Tab. 4.10. The spectra show very broad bands with absorption maxima at around 560 nm. From the edges at about 730 nm optical HOMO-LUMO gaps of about 1.70 eV can be calculated (Tab. 4.10). Compared with the monomers **BZDF9** and **10**, the absorption maxima of the polymers only show a small bathochromic shift of about 35 nm. Especially for **P-oBZDF-TH1**, the angled structure of the repeat units of the ortho-benzodifuranone chromophore is not advantageous for the planarity of the conjugated polymer chain. However, the two polymers show very broad shoulders ranging from 600 to 750 nm, which indicates an elongation of the π-conjugated system in direction of the polymer main chains.

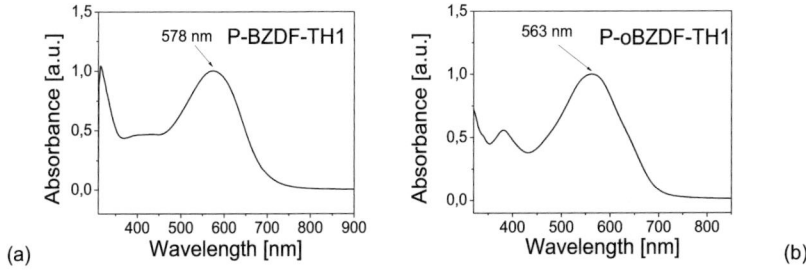

Fig. 4.25. UV-vis absorption spectra of polymers **P-BZDF-TH1** (a) and **P-oBZDF-TH1** (b).

The polymer solutions exhibit a very dark purple colour, the photographs of **P-BZDF-TH1** and **P-oBZDF-TH1** are shown in Fig. 4.27.

P-BZDF-TH1   P-oBZDF-TH1

Fig. 4.27. Photographs of **P-BZDF-TH1** and **P-oBZDF-TH1** in DCM solution.

Electrochemistry

Under similar conditions as for the polymers **P-BZDF-EDOT** and **P-oBZDF-EDOT** (see Chapter 3.1), **P-BZDF-TH1** and **P-oBZDF-TH1** were investigated using cyclic voltammetry. The cyclic voltammograms are displayed in Fig. 4.28 and the electrochemical data are complied in Tab. 4.10. Both polymers exhibit reversible oxidative and reductive cycles. **P-BZDF-TH1** shows an anodic wave at +0.56 V, which is reverted at +0.43 V, and two cathodic waves at - 0.75 and -1.10 V, which are reverted at -0.61 and -0.97 V. **P-oBZDF-TH1** exhibits a irreversible anodic wave at + 0.37 V, and an anodic wave at +0.62 V, which is reverted at +0.45 V. It also exhibits two cathodic waves at -0.78 and -1.41 V, which are reverted at -0.61 and -0.96 V. Similar to the electrochemically prepared polymers **P-BZDF-EDOT** and **P-oBZDF-EDOT**, **P-BZDF-TH1** and **P-oBZDF-TH1** also observe a stability of the anion radical reductive wave according the identical reduction mechanisms shown in Chapter 3.1.1. For both polymers, the reversibility of the oxidative and reductive cycles, and the very low HOMO-LUMO gaps of 1.00 eV for **P-BZDF-TH1**, and 0.97 eV for **P-oBZDF-TH1** suggests a strong donor-acceptor character, respectively.

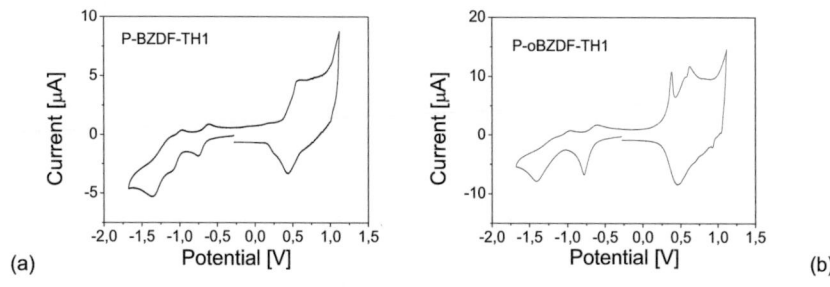

Fig. 4.28. Cyclic voltammograms of **P-BZDF-TH1** (a) and **P-oBZDF-Th1** (b) as thin films cast on a glassy carbon electrode. Solvent: 0.1 M TBAPF$_6$/acetonitrile. Potential calculated versus ferrocene. Scan rate: 100 mV s^{-1}; T = 20 °C.

Tab. 4.10. Optical and electronic data of P-**BZDF-TH1** and **P-oBZDF-TH1**.

Polymers	UV / nm	HOMO-LUMO gap (opt) / eV	Onset of oxidation / V	Onset of reduction / V	HOMO{LUMhig / eV	HOMO-LUMO gap / eV
P-BZDF-TH1	578	1.77	+0.37	-0.63	-5.17{-4.17}	1.00
P-oBZDF-TH1	562	1.78	+0.27	-0.66	-5.07{-4.14}	0.93

In summary, the synthetic route and the key properties of a series of new conjugated monomers containing benzodifuranone and thiophene units have been prepared using palladium-catalyzed Suzuki cross-coupling polycondensation reactions. The polymers exhibit broad absorptions, high HOMO- and low LUMO-levels and small band gaps (1.00 eV for **P-BZDF-TH1** and 0.97 eV for **P-oBZDF-TH1**). They also show electrochemical properties with reversible oxidation and reduction behaviour, which suggests a strong donor-acceptor character. According to their broad absorption bands, small band gaps and the reversibility of oxidation and reduction processes, the polymers might be useful for electronic applications.

4.3 Symmetrical benzodifuranone-based conjugated polymers prepared via chemical oxidative polymerization

A catalyst-free method to obtain thiophene-substituted polymers is the chemical oxidative polymerization. A strong Lewis acid, such as $FeCl_3$, is required. The synthetic route to the polymers is shown in Scheme 4.9.

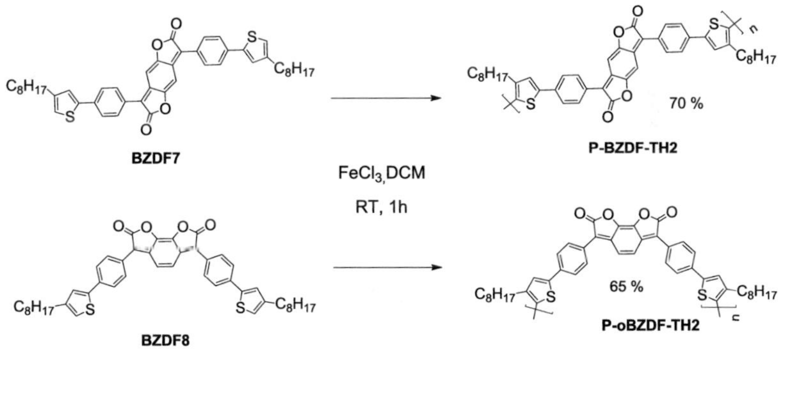

Scheme 4.9: Synthetic route of polymers **P-BZDF-TH2** and **P-BZDF-TH2**.

The chemical oxidative polymerization requires a short reaction time of 1 hour, and a dry solvent. The polymers were obtained as dark solids. They are soluble in common organic solvents. In solution they show a dark purple colour. The photographs of the polymers in DCM solution are shown in Fig. 4.29.

P-BZDF-TH2   P-oBZDF-TH2

Fig. 4.29. Photographs of **P-BZDF-TH2** and **P-BZDF-TH2** in DCM solution.

The ^{1}H-NMR spectra of **P-BZDF-TH2** and **P-oBZDF-TH2** show similar proton signals as the polymers **P-BZDF-TH1** and **P-oBZDF-TH1** in Chapter 3.2 (for NMR-spectra see Chapter 5, Experimental Part).

The ^{1}H-NMR spectra of **P-BZDF-TH2** and **P-oBZDF-TH2** display broad signals rings between 7.60 to 7.80 ppm, which can be ascribed to of the proton signals of the phenyl. The signal at about 7.40 ppm can be ascribed to the proton signal of the two thiophene units, and signals of the alkyl protons emerge in the range from 0.75 to 2.40 ppm. Most signals resemble the chemical shifts of the monomers **BZDF7** and **8**.

UV/vis absorption of the polymers

The UV/vis absorption spectra of **P-BZDF-TH2** and **P-BZDF-TH2** are displayed in Fig. 4.30, and the optical data are listed in Tab. 4.11.

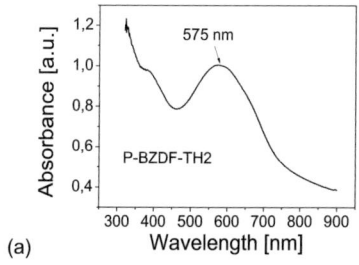 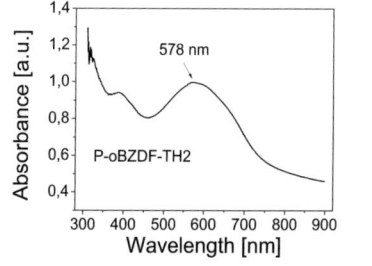

Fig. 4.30. UV-vis absorption spectra of polymers **P-BZDF-TH2** (a) and **P-oBZDF-TH2** (b).

Similar to the two benzodifuranone-based polymers **P-BZDF-TH1** and **2**, the polymers **P-BZDF-TH2** and **P-oBZDF-TH2** show broad absorption bands exhibiting maxima at about 576 nm. From the edges at about 760 nm optical HOMO-LUMO gaps of about 1.60 eV can be calculated (Tab. 4.11). Compared with the monomers **BZDF7** and **8**, the absorption maxima of the polymers exhibit a bathochromic shift of about 30 nm. However, the two polymers show very broad shoulders ranging from 650 to 800 nm, which indicates an elongation of the π-conjugated system in direction of the polymer main chains.

Electrochemistry

Under similar conditions as for the polymers **P-BZDF-TH1** and **P-oBZDF-TH1**, **P-BZDF-TH2** and **P-oBZDF-TH2** were investigated using cyclic voltammetry. The cyclic voltammograms are displayed in Fig. 4.31 and the electrochemical data are complied in Tab. 4.11. Both polymers exhibit reversible oxidative and reductive cycles. **P-BZDF-TH2** shows an anodic wave at +0.62 V, which is reverted at +0.51 V, and two cathodic waves at -0.69 and -1.05 V, which are reverted at -0.45 and -0.83 V. **P-oBZDF-TH2** exhibits an anodic wave at +0.63 V, which is reverted at +0.55 V. It also exhibits two cathodic waves at -0.74 and -1.13 V, which are reverted at -0.50 and -0.93 V. Similar to the electrochemically prepared polymers **P-BZDF-EDOT** and **P-oBZDF-EDOT**, **P-BZDF-TH1** and **P-oBZDF-TH1** also observe a stability of the anion radical reductive wave according the identical reduction mechanisms shown in Chapter 3.1.1. For both polymers, the reversibility of the oxidative and reductive cycles, and the very low HOMO-LUMO gaps of 1.00 eV for **P-BZDF-TH2**, and 0.98 eV for **P-oBZDF-TH2** suggests a strong donor-acceptor character, respectively.

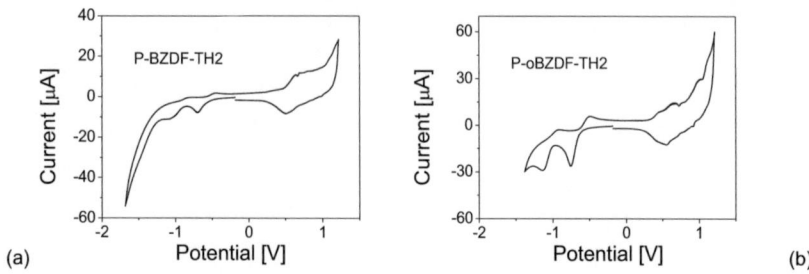

Fig. 4.31. Cyclic voltammograms of **P-BZDF-TH2** (a) and **P-oBZDF-TH2** (b) as thin films cast on a glassy carbon electrode. Solvent: 0.1 M TBAPF$_6$/acetonitrile. Potential calculated versus ferrocene. Scan rate: 100 mV s^{-1}; T = 20 °C.

Tab. 4.11. Optical and electronic data of **P-BZDF-TH2** and **P-oBZDF-TH2**.[a]

Polymers	UV / nm	HOMO-LUMO gap (opt) / eV	Onset of oxidation / V	Onset of reduction / V	HOMO{LUMO}/ eV	HOMO-LUMO gap / eV
P-BZDF-TH2	575	1.68	+0.47	-0.53	-5.27{-4.27}	1.00
P-oBZDF-TH2	578	1.60	+0.38	-0.64	-5.18{-4.20}	0.98

[a] Absorption taken in DCM solution.

In summary, a simple synthetic route to new conjugated benzodifuranone and thiophene-containing polymers have been introduced using chemical oxidative polymerization. The polymers exhibit broad absorptions, high HOMO- and low LUMO-levels and small band gaps (1.00 eV for **P-BZDF-TH2** and 0.98 eV for **P-oBZDF-TH2**). They also show electrochemical properties with reversible oxidation and reduction behaviour, which origins a strong donor-acceptor character. According to their broad absorption bands, low band gaps and the reversibility of oxidation and reduction processes, the polymers might be useful for electronic applications.

4.4 Unsymmetrical benzodifuranone-based conjugated polymers

In Chapter 3.1, it has been shown that the bifunctional symmetrical benzodifuranone monomers **BZDF 2-4** are little soluble in common solvents and not suitable for a direct palladium-catalyzed cross-coupling polycondensation such as the Suzuki cross-coupling reaction. In comparison, the unsymmetrical structure of benzodifuranone can be of great advantage in terms of solubility.

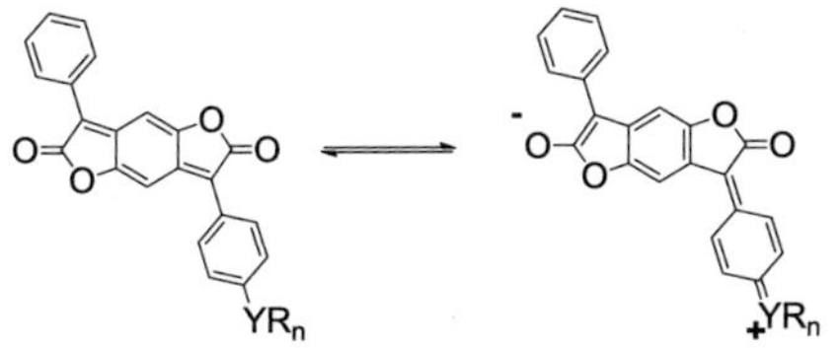

Fig. 4.32. A suggested mechanism of polarization of unsymmetrical benzodifuranone.

This could be caused by the high polarity of the molecules, which can easily form a into a zwitterionic resonance structure and interact with adjacent solvent molecules (Fig. 4.32).[62]

4.4.1 Synthesis and properties of unsymmetrical benzodifuranone monomers

The synthetic route to unsymmetrical benzodifuranone-based monomers is described in Scheme 4.10. The starting compounds **1a-b** were prepared according to literature procedures (see Chapter 5, Experimental Part). In general, the synthesis of the unsymmetrical monomers **BZDF15** and **16** required the condensation of 1 eq. hydroquinone and 1 eq. of the mandelic acid derivatives **1a** or **b**, leading to a single cyclization and formation of the intermediates 5-hydroxy-3-phenylbenzofuran-2(3H)-one (**2a** and **b**). The first ring formnation is followed by a second one with another mandelic acid derivative (**1c** and **d**) to yield the unsymmertrical benzodihydrofurofurandiones **3a** and **b**. They were oxidized to the conjugated benzodifuranone **BZDF15** and **16** with a yield of 60 to 65%. A Dean Stark apparatus was used for separation of water as byproduct favoring the cyclization reactions. The unsymmetrical benzodifuranone **BZDF15** and **16** are dark red solids. They show a very high solubility in common solvents such as chloroform, dichloromethane, DMF, THF and toluene, for example.

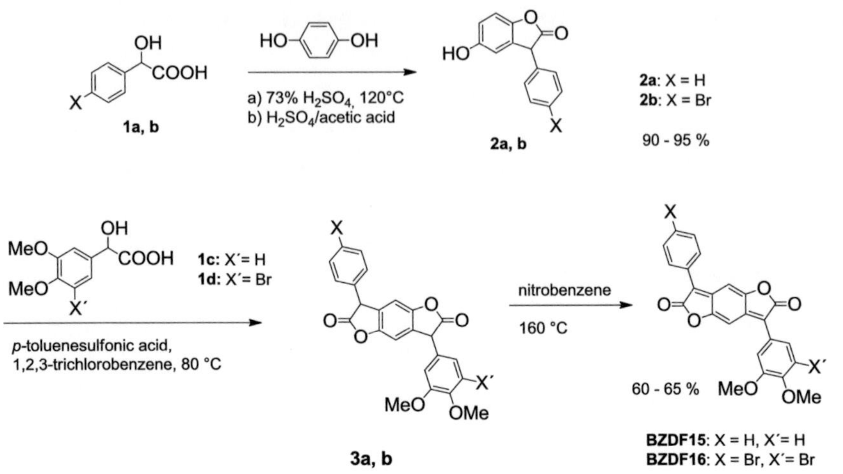

Scheme 4.10. Synthesic route to monomers **BZDF15** and **16**.

The ^{1}H-NMR spectra of **BZDF15** and **16** displayed all the expected resonances with no discernible peaks corresponding to impurities (see Chapter 5, Experimental Part). Significantly, the two adjacent singulet signals with a chemical shift at about 4.00 ppm are typical for the CH_3 groups of the benzene ring on one side of the benzodifuranone core. The other signals are similar to those of the unsubstituted monomer **BZDF1**.

UV/vis absorption and fluorescence of the monomers

The monomers **BZDF15** and **16** show a strong absorption and weak emission in the visible with large Stokes shifts in the range from 80 to 140 nm. The optical data are listed in Tab. 4.12, and the UV/vis absorption and emission spectra are displayed in Fig. 4.33. The monomers show large extinction coefficients with 42 000 and 32 000 L mol^{-1} cm^{-1}, respectively. The fluorescence quantum yields of the monomers are very low (from 0.3 to 2%). From the absorption onset of the monomers in the range from 540 to 600 nm, optical HOMO-LUMO gaps between 2.00 and 2.20 eV can be calculated. The photographs of the monomers ware shown in Fig. 4.34.

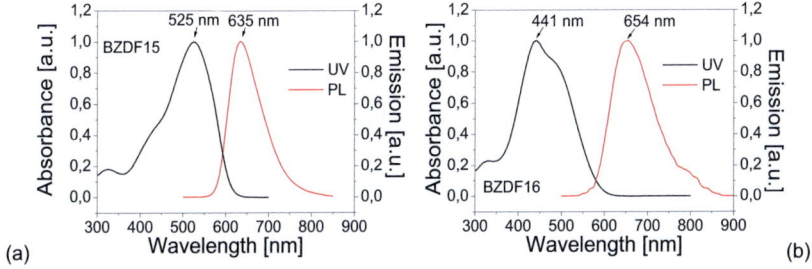

Fig. 4.33. UV/vis absorption and fluorescence of **BZDF15** (a) and **16** (b) in toluene solution.

BZDF15   **BZDF16**

Fig. 4.34. Photographs of the solution of **BZDF 15** and **16** in DCM solution.

Tab. 4.12. Optical data of monomers **BZDF15** and **16** in DCM solution.

Monomers	Absorption λ_{max} / nm	Emission λ_{max} / nm	Optical HOMO-LUMO gap / eV	Extinction coefficient ε / L mol^{-1} cm^{-1}
BZDF15	525	635	2.10	42 000
BZDF16	441	654	2.20	32 000

Solvatochromism

The polarity of the unsymmetrical monomers depends on the polarity of the solvents. Therefore, the absorption of the unsymmetrical monomers may strongly vary in solvents of different polarity. Fig. 4.35 shows the UV/vis absorption spectra of monomer **BZDF15** in different solvents, the absorption maxima vary in a range of 20 nm. In a very polar solvent, chloroform for example, the absorption maximum appears at about 535 nm, which is about 10 nm right-shifted compared to it in the less polar solvent such as acetone.

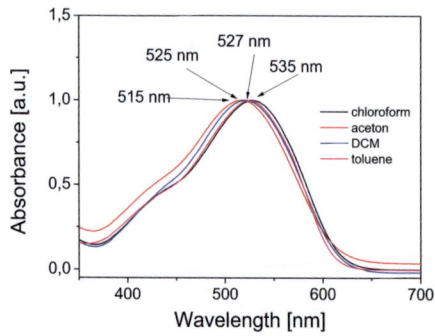

Fig. 4.35. UV/vis absorption spectra of **BZDF15** in different solvents.

4.4.2 Polymers prepared via Suzuki and Stille cross-coupling polycondensations

BZDF16 was polymerized with either 9,9-di-*n*-hexylfluorene-2,7'-bispinakolato-boronester (**FL1**) or 2,5-bis(trimethylstannyl)thiophene (**TH2**) using palladium-catalyzed cross-coupling conditions. The yields were 70% for **P-uBZDF-FL**, and 76% for **P-uBZDF-TH**, respectively. The synthetic route is described in Scheme 4.11. The polymers are dark solids, and soluble in common organic solvents.

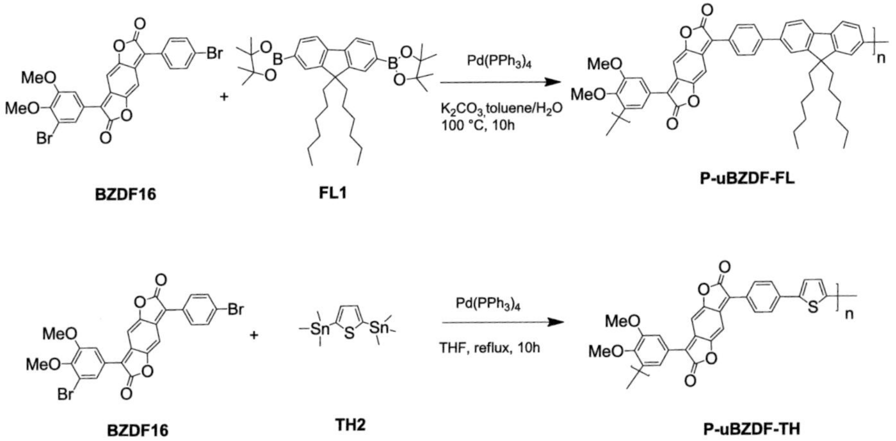

Scheme 4.11. Synthetic route of polymers **P-uBZDF-FL** and **P-uBZDF-TH**.

The ^{1}H-NMR spectra showed broad signals for aromatic protons from 6 to 8 ppm. Typical peaks at about 4 ppm can be ascribed to the signals of CH_3-protons of the methoxy-groups. For **P-uBZDF-FL**, the NMR spectrum exhibits typical proton signals of the alkyl groups attached to the fluorene unit at 0.7 - 1.6 ppm.

UV/vis absorption of the polymers

Compared to the corresponding monomer **BZDF16**, the absorption spectra of **P-uBZDF-FL** and **P-uBZDF-TH** are bathochromically shifted, resulting in large Stokes shifts of 179 nm for **P-uBZDF-FL**, and 162 nm for **P-uBZDF-TH** (Fig. 4.36). The polymers exhibit high extinction coefficients with 28 000 L mol^{-1} cm^{-1} for **P-uBZDF-FL**, and 32 500 L mol^{-1} cm^{-1} for **P-uBZDF-TH**. From the absorption onset at about 660 nm optical HOMO-LUMO gaps of 1.70 eV for **P-uBZDF-FL**, and 1.90 eV for **P-uBZDF-TH** (Tab. 4.13) can be calculated. The photographs of the polymers are shown in Fig. 4.37.

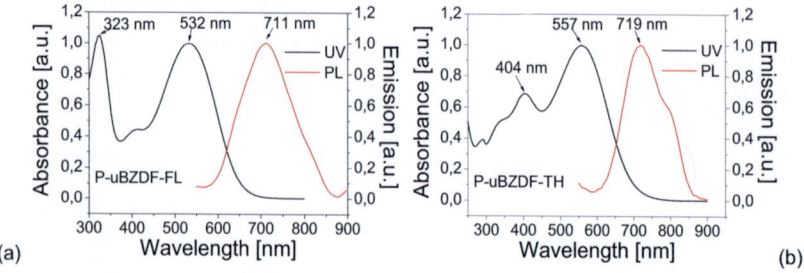

(a) (b)

Fig. 4.36. UV/Vis spectra of **P-uBZDF-FL** (a) and **P-uBZDF-TH** (b) in DCM solution.

Tab. 4.13. Optical and electronic data of **P-uBZDF-FL** and **P-uBZDF-TH**.[a]

	UV/nm	Onset of oxidation / V	HOMO / eV	Onset of reduction / V	LUMO / eV	HOMO-LUMO gap / eV
P-uBZDF-FL	323, 532	+0.26	-5.06	-0.87	-3.93	1.13
P-uBZDF-TH	404, 557	+0.29	-5.09	-0.81	-3.99	1.10

[a] Absorption spectra were taken in dichloromethane solution. All redox potentials are referenced to the ferrocene/ferrocenium redox couple. HOMO-LUMO gap according to the equation[177] $-E_{LUMO} = E_{onset(red)} + 4.8$ eV and $-E_{HOMO} = E_{onset(ox)} + 4.8$ eV, where $E_{onset(ox)}$ and $E_{onset(red)}$ are the onset potentials for the oxidation and reduction processes vs. ferrocene.

P-uBZDF-FL   P-uBZDF-TH

Fig. 4.37. Photographs of **P-uBZDF-FL** and **P-uBZDF-TH** in DCM solution.

Electrochemistry

The electrochemical response of compounds **P-uBZDF-FL** and **P-uBZDF-TH** was studied using cyclic voltammetry (Fig. 4.38). The experimental conditions are described in Chapter 5, Experimental Part. The electrochemical data of the polymers are listed in Tab. 4.13.

Both polymers show reversible oxidative and reductive waves. On increasing the potential, **P-uBZDF-FL** exhibits an anodic wave at +0.90 V. Decreasing the potential again, the oxidation is reverted at +0.57 V indicating aquasi-reversible behavior. The reductive cycle shows two cathodic waves at -0.99 and -1.52 V, which are reverted at -1.23 and -0.80 V. **P-uBZDF-TH** shows an anodic peak at +0.89 V, which is reverted at +0.50 V. Significantly, the reductive cycle shows two cathodic waves at -0.92 and -1.33 V, which are reverted at -1.19 and -0.78 V. For both polymers, a high stability of the anion radical reversible reductive wave was observed. This could possibly be explained with a stabilizing negative charge at the oxygen atoms in the carbonyl groups of the lacton groups in the benzodifuranone units. A suggested reduction mechanism is shown in Chapter 3.1.1. For both polymers **P-uBZDF-FL** and **P-uBZDF-TH**, the reversibility of the oxidative and reductive cycles, and the very low HOMO-LUMO gaps of 1.13 eV for **P-uBZDF-FL**, and 1.13 eV for **P-uBZDF-TH** suggests a strong donor-acceptor character, respectively.

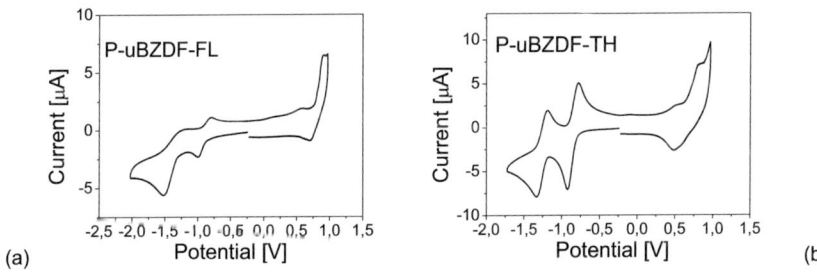

Fig. 4.38. Cyclic voltammograms of **P-uBZDF-FL** (a) and **P-uBZDF-TH** (b) as thin films on a glassy carbon electrode. Solvent: 0.1 M TBAPF$_6$/acetonitrile. Potential calculated versus ferrocene. Scan rate: 100 mV s^{-1}; T = 20 °C.

In summary, the synthetic route and the key properties of a series of new conjugated monomers containing the unsymmetrically substituted benzodifuranone unit have been presented. New π-conjugated polymers have also been synthesized using palladium-catalyzed cross-coupling methods. The polymers exhibit broad absorptions, high HOMO- and low LUMO-levels and small band gaps (1.13 eV for **P-uBZDF-FL** and 1.10 eV for **P-uBZDF-TH**). They also show interesting electrochemical properties with reversible oxidation and reduction behaviour, which suggests a strong donor-acceptor character. According to their broad absorption bands, small band gaps and the reversibility of oxidation and reduction processes, the polymers might be useful for electronic applications.

4.4.3 Polymers prepared via electropolymerization

4.4.3.1 Synthesis and properties of the monomers

A series of electropolymerizable monomers containing unsymmetrical BZDF units was also synthesized. The synthetic route is described in Scheme 4.12. The different electropolymerizable units, EDOT and bisthiophene, are attached to monomer **BZDF16** using palladium-catalyzed Stille-coupling reaction. The monomers **BZDF17** and **18** are obtained in high yield of over 80%.

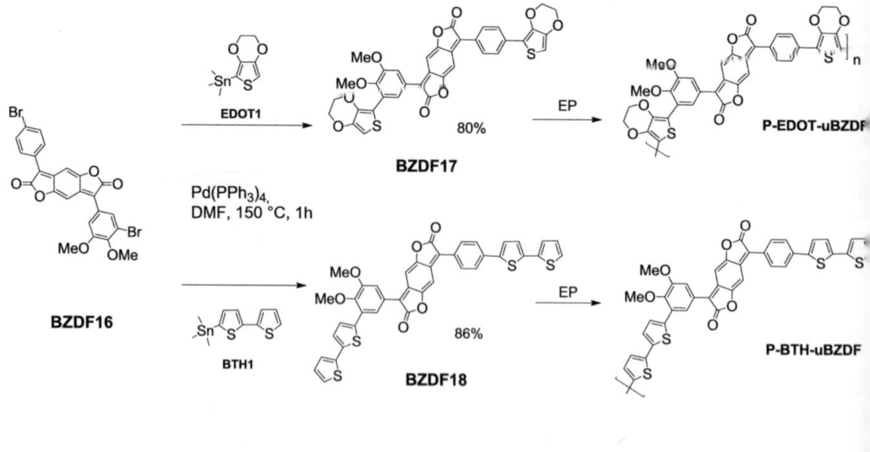

Scheme 4.12. Synthetic route to polymers **P-EDOT-uBZDF** and **P-BTH-uBZDF**.

The ¹H-NMR spectra of **BZDF17** and **18** displayed all the expected resonances with no discernible peaks corresponding to impurities (see Chapter 5, Experimental Part). The singlet proton signal at about 7.0 ppm can be ascribed to the thiophene unit in the EDOT- and dithiophene-substituent groups. Significantly, the two adjacent singlet signals with chemical shifts at about 4.0 ppm are typical for the CH_3 groups of the benzene ring on one side of the benzodifuranone core. The two triplet signals with a chemical shift of about 4.34 ppm are typical for the ethylene bridge of the EDOT unit in **BZDF17**. As for **BZDF18**, the proton signal in the range from 7.3 to 7.5 ppm can be ascribed to the aromatic protons of the thiophene units. The other signals are similar to those of the unsubstituted monomer **BZDF1**.

UV/vis absorption of the monomers

The UV/vis absorption spectra are displayed in Fig. 4.39 and the data are listed in Tab. 4.14. In the spectra of **BZDF17** and **18**, a bathochromic shift of about 22 nm is introduced by the addition of the two electropolymerizable units EDOT and bithiophene into the conjugated system of the starting compound **BZDF16**. For **BZDF18**, the large peak at 346 nm can be ascribed to the strong absorption of the large thiophene moieties in the monomer. Both monomers show very dark purple colours in solution, the photographs of **BZDF17** and **18** in DCM solution are shown in Fig. 4.40.

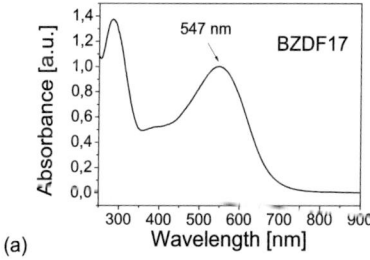
(a)

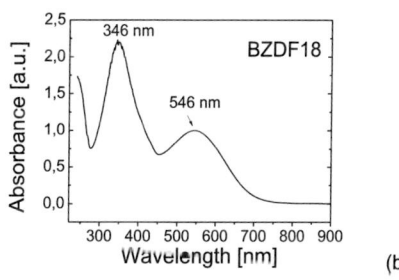
(b)

Fig. 4.39. UV/vis absorption spectra of **BZDF17** and **18** in DCM solution.

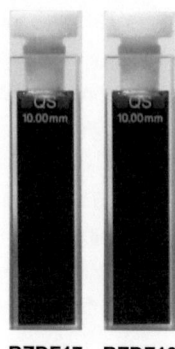

BZDF17   BZDF18

Fig. 4.40. Photographs of **BZDF17** and **18** in DCM solution.

Electrochemistry

Cyclic voltammograms of monomers

The electrochemistry of monomers **BZDF17** and **18** was studied using cyclic voltammetry in dichloromethane solution. The cyclic voltammograms of the monomers are shown in Fig. 4.41. The electrochemical data are compiled in Tab. 4.14. For the monomer **BZDF17**, the oxidative cycle exhibits irreversible peaks at +0.48 and +0.77 V. The reductive cycle shows two reversible cathodic waves at -0.95 and -1.30 V, which are reverted at -1.27 and -0.85 V. **BZDF18** shows quasi-reversible anodic peaks at +0.87 and +1.21 V, and two cathodic waves at -0.66 and -0.88 V, which are reverted at -0.84 and -0.45 V, respectively. For both monomers, the cathodic reduction shows a stable anion radical in the reduced state. Similar to the other thiophene-substituted monomers **EDOT-BZDF**, **EDOT-oBZDF**, **EDTT-BZDF**, and **EDTT-oBZDF**, the same stabilizing effect caused by the negative charges at the oxygen atoms in the carbonyl groups of both furanone rings in the bezoforofurandione monomers can also emerge. All compounds exhibit a very strong acceptor character. From the HOMO-LUMO levels, a small band gap of about 1.30 eV can be calculated (Tab. 4.14).

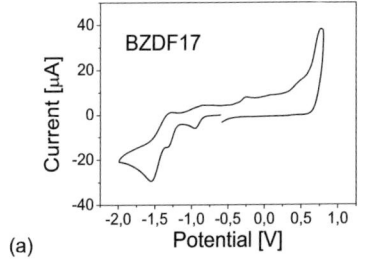
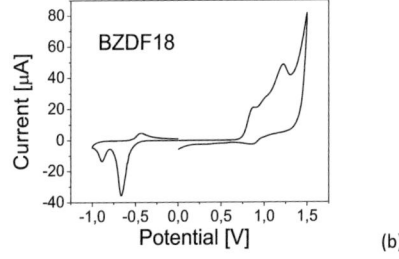

Fig. 4.41. Cyclic voltammograms of **BZDF17** (a) and **18** (b). Solvent: 0.1 M TBAPF$_6$ in DCM. Potential calculated versus ferrocene. Scan rate: 100 mV s^{-1}; T = 20 °C.

Tab. 4.14. Electrochemical data of monomers **BZDF17** and **BZDF18**. [a]

	Onset of oxidation / V	Onset of reduction / V	HOMO{LUMO} / eV	HOMO-LUMO gap / eV
BZDF17	+0.57	-0.80	-5.37{-4.00}	1.37
BZDF18	+0.75	-0.55	-5.55{-4.25}	1.30

[a]All redox potentials are referenced to the ferrocene/ferrocenium redox couple. HOMO-LUMO gap according to the equation[177] $-E_{LUMO} = E_{onset(red)}$ + 4.8 eV and $-E_{HOMO} = E_{onset(ox)}$ + 4.8 eV, where $E_{onset(ox)}$ and $E_{onset(red)}$ are the onset potentials for the oxidation and reduction processes vs. ferrocene.

4.4.3.2 Polymers prepared via electropolymerization

Electropolymerization

The two monomers were electropolymerized using the same conditions as described in Chapter 3.1. The polymer growth plots are shown in Fig. 4.42. Both monomers polymerized readily by continuously scanning between -0.2 and +1.0 V. New peaks appeared in lower voltage areas. This indicates a lower energy level of the resulting polymers. The elongation of the conjugated system emerges during the polymerization. Significantly, the bithiophene-substituted **P-BTH-uBZDF** grew better than the EDOT-substituted **P-EDOT-uBZDF**. This can be ascribed to polymerization of the large thiophene moieties in **BZDF18**, which are advantagous for electropolymerization.

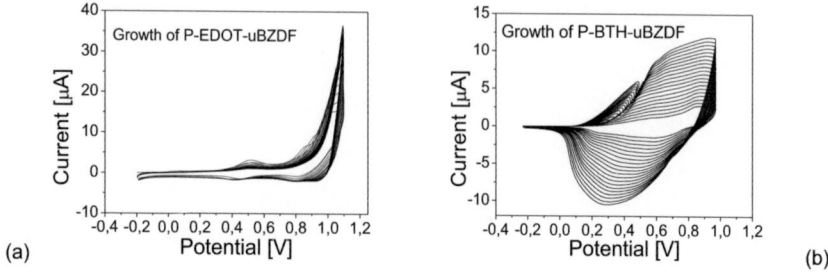

Fig. 4.42. Growth of **P-EDOT-uBZDF** (a) and **P-BTH-uBZDF** (b) by cyclic voltammetry in dichloromethane using a carbon working electrode, an Ag wire aching as pseudo-reference electrode. Supporting electrolyte: 0.1 M TBAPF$_6$. Scan rate: 100 mV s^{-1}; T = 20 °C.

UV/vis absorption of polymers

The UV/vis absorption spectra of the two polymers are shown in Fig. 4.43, and the data are listed in Tab. 4.15. They show very broad bands with absorption maxima at around 600 nm for **P-EDOT-uBZDF** and 554 nm for **P-BTH-uBZDF**. From the edge at about 700 nm for **P-BTH-uBZDF** an optical HOMO-LUMO gap of about 1.75 eV can be calculated (Tab. 4.15). The absorption edge of **P-EDOT-uBZDF** is due to the broad absorption shoulder incalculable.

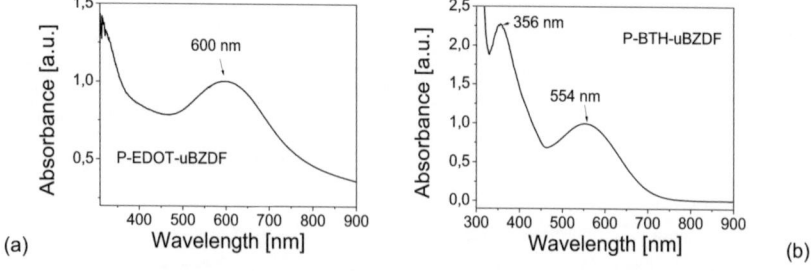

Fig. 4.43. UV/vis absorption spectra of **P-EDOT-uBZDF** (a) and **P-BTH-uBZDF** (b) in films.

Compared with the monomers **BZDF17**, **P-EDOT-uBZDF** exhibits a bathochromical shift of 53 nm. In comparison, **P-BTH-uBZDF** only exhibits a small bathochromical shift of 10 nm compared with the corresponding monomer **BZDF18**. The reason could be that the same as for the EDOT-substituted polymers **P-EDOT-BZDF**, in which the polymer main chain exhibits a better planarity due to the interaction between the oxygen and sulfur atoms of the adjacent monomer units. The main chain of **P-BTH-uBZDF** does not provide this planarity, therefore, it may be strongly twisted. This may interrupt the planarity and the π-conjugation of the conjugated polymer chain. However, the very broad shoulders ranging from 600 to 750 nm indicate an elongation of the π-conjugated system in direction of the polymer main chains.

Cyclic voltammograms of polymers

The CV diagrams of **P-EDOT-uBZDF** and **P-BTH-uBZDF-TH** are shown in Fig. 4.44. For **P-EDOT-uBZDF**, anodic oxidation leads to a quasi-reversible peak at +0.90 V. Significantly, the reductive cycle shows two cathodic waves at -0.76 and -1.20 V, which are reverted at -1.0 and -0.57 V. **P-BTH-uBZDF-TH** shows an anodic wave at +0.73 V, which is reverted at +0.35 V, and two reversible cathodic waves at -1.00 and -1.43 V, which are reverted at -1.26 and -0.85 V. It is to note, that a high stability of both cation and anion radical was observed. With small HOMO-LUMO gaps of 1.22 eV for **P-EDOT-uBZDF** and 1.13 eV for **P-BTH-uBZDF-TH**, a strong donoer-acceptor character is likely.

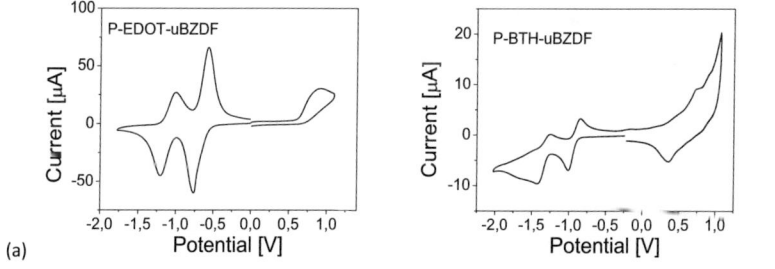

Fig. 4.44. Cyclic voltammograms of **P-EDOT-uBZDF** (a) and **P-BTH-uBZDF-TH** (b). Solvent: 0.1 M TBAPF$_6$ in DCM. Potential calculated versus ferrocene. Scan rate: 100 mV s^{-1}; T = 20 °C.

Tab. 4.15. Optical and electrochemical data of **P-EDOT-uBZDF** and **P-BTH-BZDF** as thin films.

	UV/nm	HOMO-LUMO gap (opt) / eV	Onset of oxidation / V	Onset of reduction / V	HOMO {LUMO} / eV	HOMO-LUMO gap / eV
P-EDOT-uBZDF	600	-	+0.63	-0.59	-5.43{-4.21}	1.22
P-BTH-uBZDF	554	1.75	+0.24	-0.89	-5.04{-3.91}	1.13

In summary, new π-conjugated polymers have been synthesized by electropolymerization under anodic conditions. The polymers exhibit broad absorptions, high HOMO- and low LUMO-levels, and small band gaps. Similar to the polymers **P-uBZDF-FL** and **P-uBZDF-TH** prepared by palladium-catalyzed polycondensation reactions, **P-EDOT-uBZDF** and **P-BTH-uBZDF** also show stability and reversibility of oxidation and reduction processes. This renders the compounds useful for electronic applications.

4.5 Conclusion

In this chapter, a synthetic method of bifunctional benzodifuranone monomers was shown, and the first incorporation of benzodifuranone chromophore into conjugated polymer chains was successfully introduced.

Conjugated polymers based on symmetrical and unsymmetrical benzodifuranones were prepared via palladium-catalyzed Suzuki polycondensation reactions, or by using electropolymerization under cathodic conditions, in which the benzodifuranone chromophore was attached to electropolymerizable units such as EDOT and EDTT, or by a palladium-free method, the chemical oxidative polymerization.

The polymers show very broad absorption bands with maxima up to 600 nm, exhibiting extinction coefficients in the range from 25 000 to 60 000 L mol^{-1} cm^{-1}. Significantly, the polymers exhibit reversible anodic and cathodic behaviours, which shows a strong electron donor and acceptor character. The polymers show low HOMO-LUMO band gaps in the range from 0.40 to 1.20 eV. Interesting electrochromic properties were found under cathodic conditions. Under reduction, the polymers showed colour changes from dark to almost transparent.

Due to their broad absorption bands, low band gaps and reversibility of oxidation and reduction processes, the polymers might be useful for optical and electronic application such as organic solar cells or electrochromic devices.

5 Experimental part

5.1 Materials

Chemicals for preparative and analytical procedures were purchased from the chemical suppliers Acros, Alfa Aeser, and Sigmal-Aldrich and used without further purification, if not otherwise stated. DPP pigments were kindly provided by BASF Schweiz AG, Basel, Switzerland. The purification of solvents such as acetone, DCM, chloroform, methanol, ethanol, DMF, DMSO, THF and toluene was carried out according to literature procedures.[187]

5.2 Instrumentation and general procedures

^{1}H-NMR spectra were recorded on a Bruker DPX instrument at 300, 400 or 500 MHz, and ^{13}C-NMR spectra at 100 MHz. Chemical shifts are given in parts per million. Deuterated solvents such as chloroform-d and DMSO-d were used for the sample preparation.

UV/vis absorption spectra were recorded on a Perkin-Elmer Lambda 14 spectrometer using quartz cuvettes (thickness 1 cm) and solvents of spectroscopic grade. Samples in the solid state were measured on quartz substrates.

Photoluminescence spectra were recorded on a Perkin-Elmer LS 50 B spectrometer under the same conditions as for the UV/vis measurements.

IR spectra were recorded on a Perkin-Elmer FTIR spectrometer.

Elemental analyses were obtained using a Perkin-Elmer 2400 elemental analyzer.

Electrochemical measurements were performed on a Heka potentiostat PG390 or CH Instruments 660A electrochemical workstation with IR compensation using anhydrous DCM or acetonitrile as the solvent, silver wire as the pseudo reference electrode, and platinum wire and glassy carbon as the counter and working electrodes, respectively. Electrochemical data were refered to the ferrocene/ferrocenium redox couple using the metallocene as an internal standard. All solutions were purged with Ar. The monomer concentrations were ca. 1×10^{-3} M, n-Bu$_4$NPF$_6$ (0.1 M) was used as the supporting electrolyte. Spectroelectrochemical experiments were conducted on ITO glass electrodes.

The gel electrolyte for electrochemical measurements was prepared according to a literature procedure.[188] The electrolyte contains 70 wt.% of acetonitrile, 20 wt.% of propylene carbonate, 7 wt.% of PMMA (Mw: 50 000 Da), and 3 wt.% of tetrabutylammonium hexafluorophosphate.

Air- and/or water-sensitive reactions were conducted under nitrogen using dry solvents. Microwave assisted syntheses were carried out using a Biotage Initiator Sixty EXP Microwave System.

Molecular weights were determined by size exclusion chromatography (SEC) using a Waters/Millipore UV detector 481 and an SEC column combination (Latek/styragel 50/1000 nm pore size). All measurements were carried out in tetrahydrofuran at 45 °C. The columns were calibrated versus commercially available polystyrene standards.

The porous properties of the networks were investigated by nitrogen adsorption and desorption at 77.3 K using an ASAP2020 volumetric adsorption analyzer (Micrometrics Instrument Corporation). Samples were degassed at 110 °C for 15 h under vacuum (10-5 bar) before analysis. Pore structure properties of the samples were determined via nitrogen adsorption and desorption at -196 °C using a volumetric technique on an ASAP2020 adsorption analyzer (Micromeritics Instrument Corporation). Before analysis, the samples were degassed at 110 °C over night under vacuum (10-5 Bar). Brunauer-Emmett-Teller (BET) surface area was obtained in the relative pressure (p/p_0) range from 0.05 to 0.20, and the total pore volume was determined from the amount of nitrogen adsorbed at p/p_0 = ca. 0.99.

High resolution imaging of the polymer morphology was achieved using a Zeiss NEON 40 Scanning Electron Microscope (SEM). Samples were coated with a layer of gold using an Emitech K550X automated sputter coater before analysis.

5.3 Synthesis

5.3.1 Diketopyrrolopyrrole-based conjugated polymers

3,6-Bis(4-(2,3-dihydrothieno[3,4-b][1,4]dioxin-5-yl)phenyl)-2,5-bis(2-hexylundecyl)-pyrrolo[3,4-c]pyrrole-1,4(2H,5H)-dione (EDOT-DPP1a)

In a vial 200 mg (0.22 mmol) 3,6-bis(4-bromophenyl)-2-(2,5-bis(2-hexyldecyl)pyrrolo[3,4-c]pyrrole-1,4(2H,5H)-dione **(HDBrDPP)**, 168 mg (0.55 mmol) (2,3-dihydro- thieno[3,4-b][1,4]dioxin-5-yl) trimethylstannane and 12.71 mg (0.011 mmol) tetrakis-(triphenylphosphine) palladium(0) were dissolved in 5 ml dry DMF and stirred for 5 min. The mixture was degassed, and heated at 160 °C for 1 h in the microwave synthesizer. After cooling the mixture was diluted with 50 ml DCM, and washed with 50 ml brine and 50 ml water. The organic layer was separated, dried over magnesium sulfate and evaporated. The red product was recrystallised from DCM/methanol. Yield: 196 mg (90%). UV/vis (DCM): λ_{max} = 500, 377, 320 nm. Fluorescence (DCM): λ_{max} = 531 nm. ^{1}H-NMR (300 MHz, CDCl$_3$): δ = 7.84 (d, DPP phenylene, 8H), 6.38 (s, EDOT aromatic H H, 2H), 4.33 (t, EDOT-CH$_2$, 8H), 3.82 (t, N-CH$_2$, 4H), 1.57 (m, CH, 2H), 1.14 (m, CH$_2$, 56H), 0.85 (t, CH$_3$, 12H).

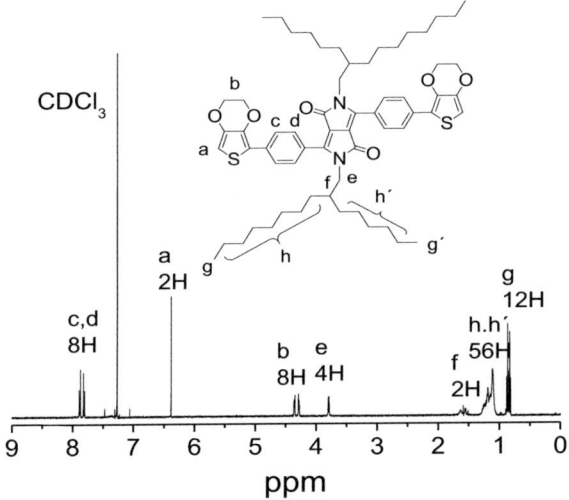

2,5-Bis(4-tert-butylphenyl)-3,6-bis(4-(2,3-dihydrothieno[3,4-b][1,4]dioxin-5-yl)phenyl)-pyrrolo[3,4-c]pyrrole-1,4(2H,5H)-dione (EDOT-DPP1b)

The same procedure was used as for compound **EDOTDPP1a**. Recrystallization from DCM/methanol yielded **EDOTDPP1b** as a dark red powder (85%). UV/vis: λ_{max} = 532, 394, 320 nm. ^{1}H-NMR (300 MHz, CDCl$_3$): δ = 7.69, 7.42, 7.17 (d, DPP phenylene, 16H), 6.38 (s, EDOT aromatic H H, 2H), 4.26 (t, EDOT-CH$_2$, 8H), 1.35 (s, CH$_3$, 18H).

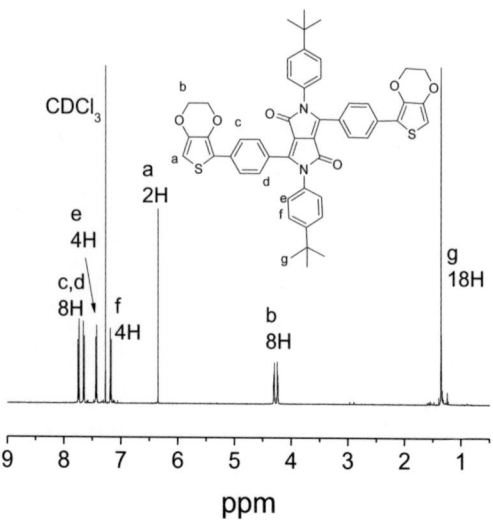

3,6-Bis(4-tert-butylphenyl)-2,5-bis(4-(2,3-dihydrothieno[3,4-b][1,4]dioxin-5-yl)phenyl)-pyrrolo[3,4-c]pyrrole-1,4(2H,5H)-dione (EDOT-DPP2)

The same procedure was used as for compound **EDOT-DPP1a** and **1b**. Recrystallisation from DCM/methanol yielded **EDOT-DPP2** as a dark red powder (82%). UV/vis: λ_{max} = 488 nm, 311 nm. ^{1}H-NMR (300 MHz, CDCl$_3$): δ = 7.76, 7.36, 7.22 (m, DPP phenylene, 16H), 6.36 (s, EDOT aromatic H H, 2H), 4.31 (t, EDOT-CH$_2$, 8H), 1.58 (s, CH$_3$, 18H).

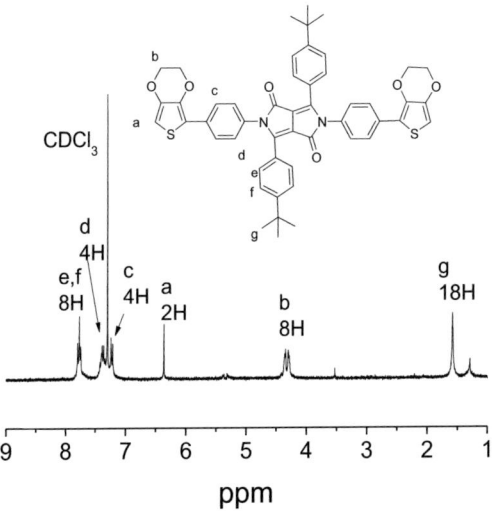

2,3,5,6-Tetrakis(4-bromophenyl)pyrrolo[3,4-c]pyrrole-1,4(2H,5H)-dione (*t*-BrDPP)

3,6-Bis(4-bromophenyl)furo[3,4-c]furan-1,4-dione (**BrDFF**)[13] (500 mg, 1.12 mmol), *p*-bromoaniline (576 mg, 3.36 mmol), dicyclohexylcarbodiimide (692 mg, 3.36 mmol) and trifluoroacetic acid (6 μl, 0.07 mmol) were dissolved in 250 ml chloroform and stirred for 3 d at room temperature. The solvent was removed, and the residue washed with methanol. After filtration, the red coloured product was purified by column chromatography (toluene). The product was recrystallized from methanol as red crystals with a red solid state

149

fluorescence. Yield: 466 mg (55%). ¹H-NMR (400 MHz, CDCl₃): δ = 7.56 (d, aromatic H, 4H), 7.51 (d, aromatic H, 8H), 7.07 (d, aromatic H, 4H). ¹³C NMR (100 MHz, CDCl₃): δ = 132.6, 132.0, 131.0, 129.1. UV/Vis (DCM): λ_{max} = 338 nm, 477 nm, 503 nm. Mp > 300 °C.

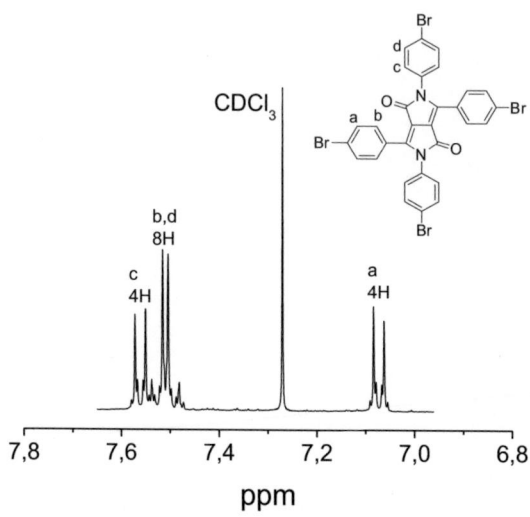

2,3,5,6-Tetrakis(4-(4-hexylthiophen-2-yl)phenyl)pyrrolo[3,4-c]pyrrole-1,4(2H,5H)-dione (*t*-DPP1)

In a vial, 150 mg (0.20 mmol) 2,3,5,6-tetrakis(4-bromophenyl)pyrrolo[3,4-c]pyrrole-1,4(2H,5H)-dione (**5**), 328 mg (1.00 mmol) 4-hexylthien-2-yl trimethylstannane, and 14 mg

(0.012 mmol) tetrakis(triphenylphosphine) palladium(0) were dissolved in 5 ml dry DMF, and stirred for 5 min. The mixture was degassed, and heated at 160 °C for 1 h in the microwave synthesizer. After cooling, the mixture was diluted with 75 ml DCM, and washed with 100 ml brine and 100 ml water. The organic layer was separated, dried over magnesium sulfate and evaporated. The red solid was recrystallized from DCM/methanol. Yield: 190 mg (86%). ^{1}H-NMR (400 MHz, CDCl$_3$): δ = 7.75 (d, aromatic H, 4H), 7.65 (d, aromatic H, 4H), 7.46 (d, aromatic H, 4H), 7.24 (s, aromatic H, thienyl-, 4H), 7.20 (d, aromatic H, 4H), 6.91 (s, aromatic H, thienyl-, 4H), 6.22 (q, CH$_2$-H, 8H), 1.64 (m, CH$_2$-H, 24 H), 1.34 (m, CH$_2$-H, 8H), 0.98 (t, CH$_3$-H, 12H). UV/Vis (DCM): λ$_{max}$ = 309, 395, 520 nm. ε (520) = 7.3 · 10^3 L mol^{-1}cm^{-1}.

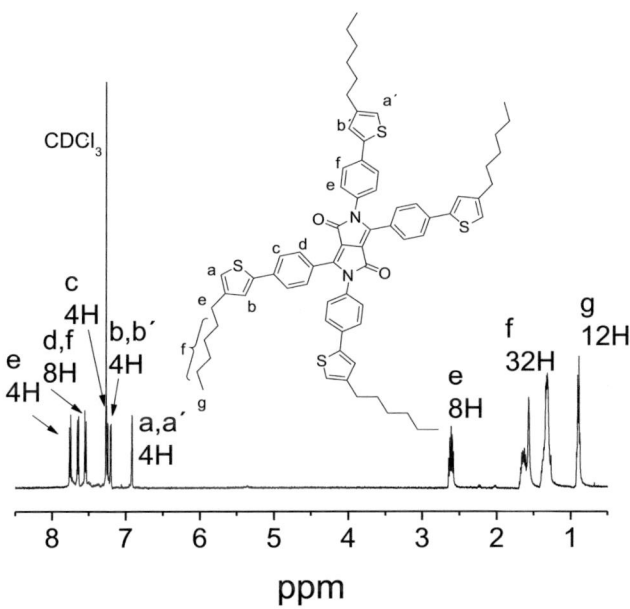

2,3,5,6-Tetrakis(4-(2,3-dihydrothieno[3,4-b][1,4]dioxin-5-yl)phenyl)pyrrolo[3,4-c]pyrrole-1,4(2H,5H)-dione (*t*-DPP2)

The procedure from compound **t-DPP1** was followed except that 3,4-ethylenedioxythien-2-yl trimethylstannane was used instead of 4-hexylthien-2-yl trimethylstannane, giving a dark red solid in a yield of 79%. ^{1}H-NMR (400 MHz, CDCl$_3$): δ = 7.84 (d, aromatic H, 8H), 6.38 (s, EDOT aromatic H H, 2H), 4.33 (t, EDOT-CH$_2$, 8H), 3.82 (t, N-CH$_2$, 4H), 1.57 (m, CH, 2H), 1.14 (m, CH$_2$, 56H), 0.85 (t, CH$_3$, 12H). UV/Vis (DCM): λ$_{max}$ = 317, 535 nm. ε (535) = 7.7 · 10^3 L mol^{-1}cm^{-1}.

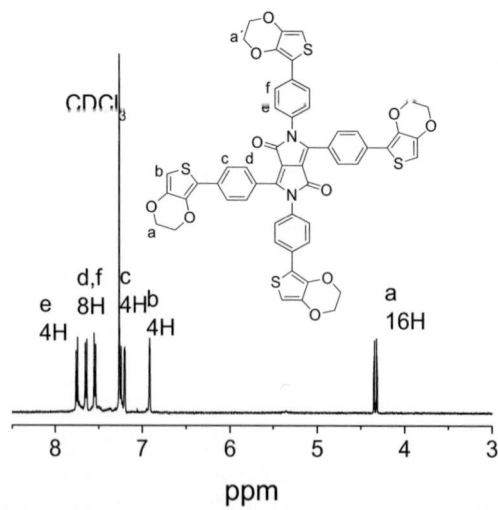

2,3,5,6-Tetrakis(4-(2,3-dihydrothieno[3,4-b][1,4]dithiin-5-yl)phenyl)pyrrolo[3,4-c]pyrrole -1,4(2H,5H)-dione (*t*-DPP3).

The procedure from ***t*-DPP1** was followed except that 3,4-ethylenedithiathien-2-yl trimethylstannane was used instead of 4-hexylthien-2-yl trimethylstannane, giving a dark solid in a yield of 81%. ^{1}H-NMR (400 MHz, CDCl$_3$): δ = 7.85 (d, aromatic H, 8H), 6.68 (s, EDTT aromatic H H, 2H), 3.24 (t, EDTT-CH$_2$, 8H), 3.80 (t, N-CH$_2$, 4H), 1.57 (m, CH, 2H), 1.14 (m, CH$_2$, 56H), 0.88 (t, CH$_3$, 12H). UV/Vis (DCM): λ$_{max}$ = 253, 309, 519 nm. ε (519) = 5.3 · 10^3 L mol^{-1}cm^{-1}.

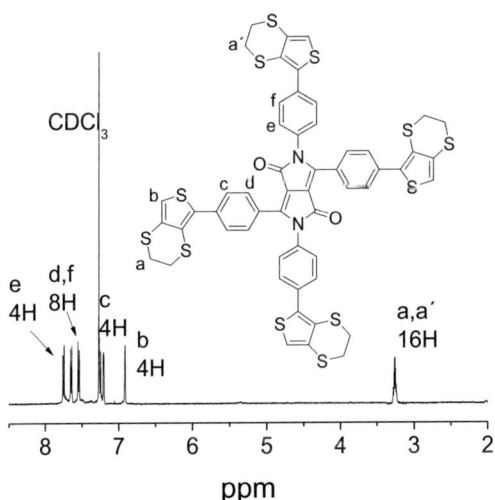

Network N1

In a vial, 2,3,5,6-tetrakis(4-bromophenyl)pyrrolo[3,4-c]pyrrole-1,4(2H,5H)-dione (**t-BrDPP**) (200 mg, 0.26 mmol), Ni(COD)$_2$ (343 mg, 1.25 mmol), 2,2'-dipyridyl (183 mg, 1.25 mmol) and cyclooctadiene (135 mg, 1.25 mmol) were dissolved in 10 ml dry DMF. The mixture was degassed and heated under nitrogen at 100 °C for 2 h in the microwave synthesizer. After cooling, the precipitated solid was filtered off, and washed with acetone, DCM, 1 M HCl, and water until acid-free. The red solid was dried under vacuum yielding 90 mg (83%). The solid was completely insoluble in common organic solvents. HR-MAS-NMR: (500 MHz, CDCl$_3$): δ = 7.75 - 7.05 (m, aromatic H). Anal. Calculated for C$_{30}$H$_{16}$N$_2$O$_2$: C, 82.56%; H, 3.69%; N 6.42%. Found: C, 81.24%; H, 4.60%; N, 5.45%. FT-IR (cm^{-1}): 1688 (C=O), 1603 and 1492 (C=C).

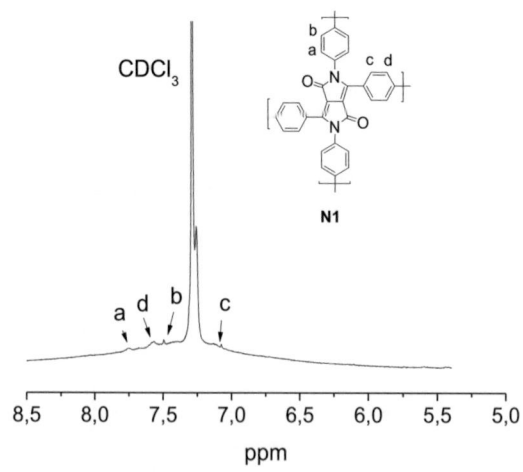

Network N2

In a vial, 2,3,5,6-tetrakis(4-bromophenyl)pyrrolo[3,4-c]pyrrole-1,4(2H,5H)-dione (*t*-**BrDPP**) (200 mg, 0.26 mmol), 1,4-diethynylbenzene (66 mg, 0.52 mmol), tetrakis(triphenylphosphine) palladium(0) (18 mg, 0.016 mmol), copper(I) bromide (2.3 mg, 0.016 mmol) were dissolved in a mixture of DMF (5 ml) and Et$_2$N. The reaction mixture was degassed and heated under nitrogen at 100 °C for 2 h in the microwave synthesizer. The red solid precipitate was filtered off, and washed with acetone, DCM, 1 M HCl, and water until acid-free. The product was dried under vacuum yielding 128 mg (82%). The solid was absolutely insoluble in common organic solvent. HR-MAS-NMR: (500 MHz, CDCl$_3$): δ = 7.65 - 7.10 (m, aromatic H). FT-IR (cm^{-1}): 2205 (C≡C), 1696 (C=O), 1599 and 1513 (C=C).

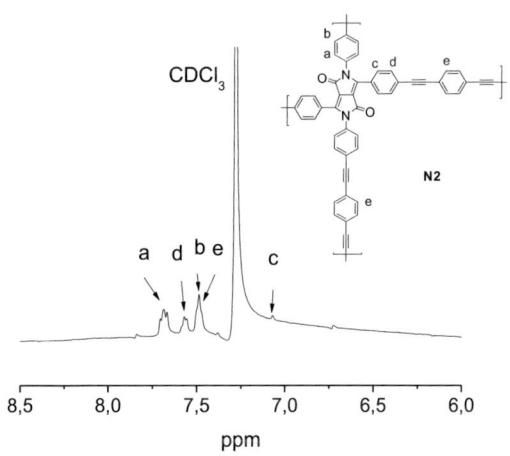

Network N3

The procedure from **N2** was followed except that 4,4'-diethynylbiphenyl was used instead of 1,4-diethynylbenzene, giving a red solid in a yield of 79%. The solid was completely insoluble in common organic solvents. HR-MAS-NMR: (500 MHz, CDCl$_3$): δ = 7.65 - 7.10 (m, aromatic H). FT-IR (cm^{-1}): 2210 (C≡C), 1687 (C=O), 1599 and 1507 (C=C).

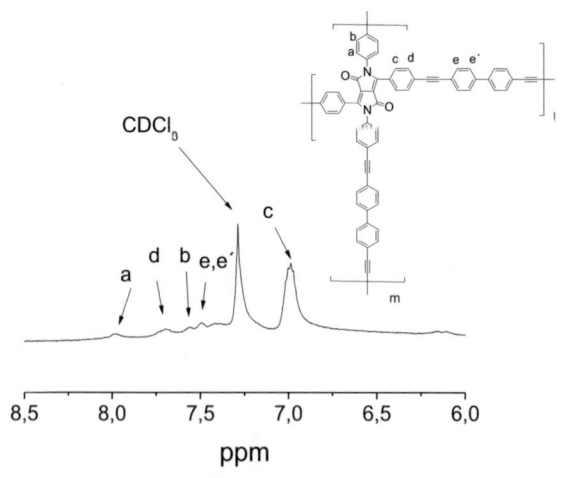

Network N4

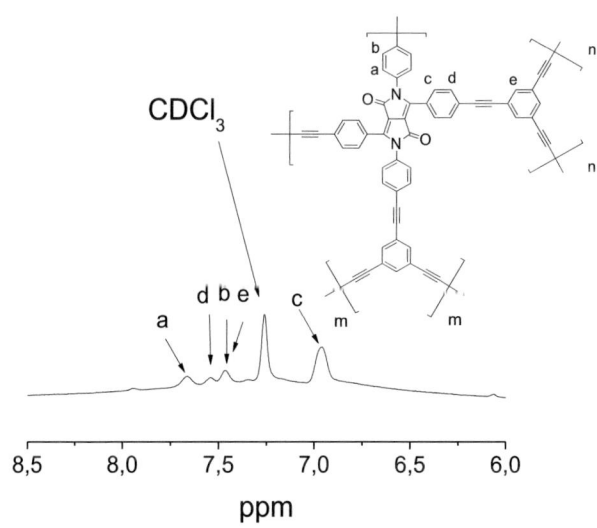

The procedure from **N2** was followed except that 1,3,5-triethynylbenzene was used instead of 1,4-diethynylbenzene, giving a red solid in a yield of 80%. The solid was completely insoluble in common organic solvents tested. HR-MAS-NMR: (500 MHz, CDCl$_3$): δ = 7.65 - 6.90 (m, aromatic H). FT-IR (cm^{-1}): 2203 (C≡C), 1694 (C=O), 1583 and 1505 (C=C).

Poly(DPP-phenylene) (P-DPP1)

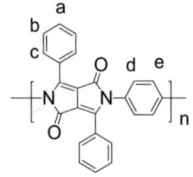

In a 20 ml flask 3,6-diphenylfuro[3,4-c]furan-1,4-dione (**DFF**) (100 mg, 0.35 mmol), *p*-phenylenediamine (37 mg, 0.35 mmol), and *N,N'*-dicyclohexylcarbodiimide (235 mg, 1.14 mmol) were dissolved in 10 ml chloroform. 2.6 μl trifluoroacetic acid were dropped in the reaction mixture using a syringe. The solution was stirred at room temperature for 3 days. Then the solvent was evaporated, and the precipitated crude product was washed with 20 ml methanol and dried in air. Yield: 64 mg (51%). ^{1}H-NMR (300 MHz, DMSO-d): δ = 8.34 (m, aromatic H), 7.72 - 6.41 (m, aromatic H). UV/vis (DCM): λ_{max} = 461 nm. Molecular weight: 3500 Da. Polydispersity: 1.6.

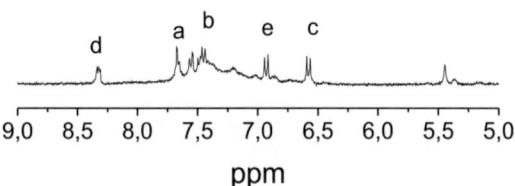

Poly(DPP-dimethoxydiphenyl) (P-DPP2)

The same procedure as for **P-DPP1** was used except that p-phenylenediamine was replaced by 3,3'-dimethoxy-[1,1'-biphenyl]-4,4'-diamine. A red solid was obtained in the yield of 60%. ^{1}H-NMR (300 MHz, DMSO-d): δ = 8.43 - 6.10 (m, aromatic H), 4.05 - 3.50 (m, OMe-H). UV/vis (DCM): λ_{max} = 453 nm. Molecular weight: 4500 Da. Polydispersity: 2.1.

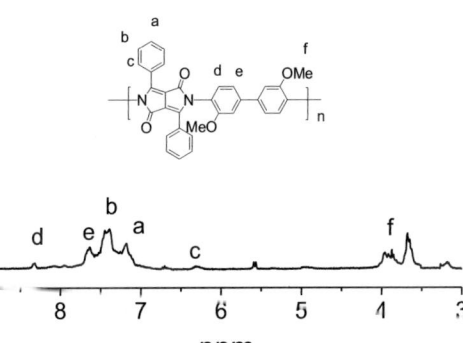

Poly(DPP-diphenylmethylene) (P-DPP3)

The same procedure as for **P-DPP1** was used except that p-phenylenediamine was replaced by 4,4'-methylenedianiline. A red solid was obtained in the yield of 60%. ^{1}H-NMR (300 MHz, DMSO-d): δ = 7.80 - 6.50 (m, aromatic H), 3.50 (m, CH_2-H). UV/vis (DCM): λ_{max} = 472, 491 nm. Molecular weight: 4500 Da. Polydispersity: 2.1.

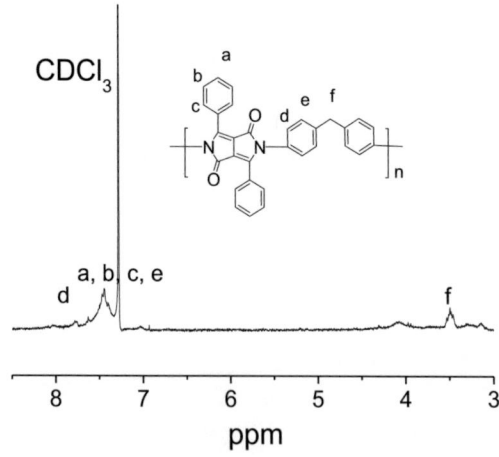

Poly(DPP-dodecane) (P-DPP4)

The same procedure as for **P-DPP1** was used except that *p*-phenylenediamine was replaced by or dodecane-1,12-diamine. A yellow solid was obtained in the yield of 60%. ^{1}H-NMR (300 MHz, CDCl$_3$): δ = 7.86 - 6.15 (m, aromatic H), 4.12 - 3.40 (m, N-CH$_2$-H), 2.02 - 1.03 (m, CH$_2$-H). UV/vis (DCM): λ_{max} = 473, 495 nm. Molecular weight: 4100 Da. Polydispersity: 2.6.

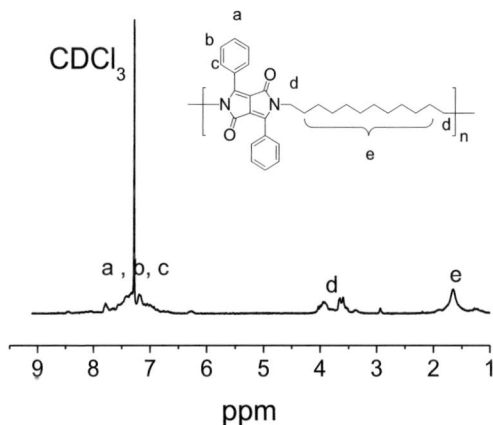

3,6-Di(thiophen-2-yl)pyrrolo[3,4-c]pyrrole-1,4(2H,5H)-dione (ThDPP)[189]

A flask was charge with potassium *tert*-butylate (8 g, 71.4 mmol) was added to a round flask with argon protection. Then a solution of *t*-amyl alcohol (50 mL) and 2-thiophenecarbonitrile (6.50 g, 60 mmol) was injected by a syringe one portion. The mixture was warmed up to 100−110 °C, and a solution of dimethyl succinate (2.92 g, 20 mmol) in *t*-amyl alcohol (10 mL) was dropped slowly in 1 h. After the addition, the reaction mixture was stirred at the same temperature for about 3 h. Then the mixture was cooled to 65 °C, diluted with 50 mL of methanol, and neutralized with acetic acid and reflux for another 10 min. Then the suspension is filtered, and the black filter cakewas washed by hot methanol and water twice each and dried in vacuum yielding the crude product (5 g, 84%), which could be used directly to next step without further purification.

Di-tert-butyl-1,4-dioxo-3,6-di(thiophen-2-yl)pyrrolo[3,4-c]pyrrole-2,5(1H,4H)-dicarboxylate (Boc-ThDPP)[22]

To a suspension of 3,6-di(thiophen-2-yl)pyrrolo[3,4-c]pyrrole-1,4(2H,5H)-dione (**ThDPP**) (1.3 g, 4.33 mmol) in 100 ml of dried THF were added 4-dimethylaminopyridine (DMAP) (0.26 g, 2.16 mmol) and di-*tert*-butyldicarbonate (4.72 g, 21.6 mmol). The reaction mixture was stirred at room temperature with the exclusion of atmospheric moisture over night. Subsequently, the reaction mixture is poured under good stirring into 500 ml of distilled water. The precipitated brown solid is isolated by filtration, the residue is washed with cold water and dried under vacuum at room temperature. The crude product was dissolved in 100 mL dichloromethane and filtered through celite. The solution was washed with 50 ml water, and dried over anhydrous MgSO$_4$. Then the solvent was evaporated and the product was recrystallized from DCM/methanol, yielding 1.92 g (89%) product. UV/vis: λ_{max} = 495 nm. ^{1}H-NMR (300 MHz, CDCl$_3$): δ = 8.25 (d, aromatic H, 2H), 7.65 (d, aromatic H, 2H), 7.22 (t, aromatic H, 2H), 1.61 (s, CH$_3$, 18H).

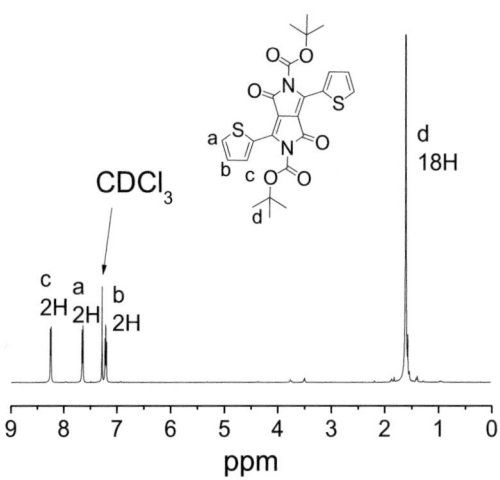

Di-tert-butyl-3,6-bis(5-bromothiophen-2-yl)-1,4-dioxopyrrolo[3,4-c]pyrrole-2,5(1H,4H)-dicarboxylate (BrBoc-ThDPP)

In a 250 ml flask, di-tert-butyl-1,4-dioxo-3,6-di(thiophen-2-yl)pyrrolo[3,4-c]pyrrole-2,5(1H,4H)-dicarboxylate (**Boc-ThDPP**) (3.7 g, 7.4 mmol) was dissolved in 150 ml chloroform, and Nbromosuccinimide (NBS) (1.9 g, 16.64 mmol) was then in one portion at room temperature. The mixture was stirred over night. Then the solution was washed with 100 ml water, and dried over MgSO$_4$, and the solvent was evaporated. The product was recrystallized from DCM/methanol, yielding 4.38 g (90%). UV/vis: λ_{max} = 515 nm. ^{1}H-NMR (300 MHz, CDCl$_3$): δ = 8.10 (d, aromatic H, 2H), 7.17 (d, aromatic H, 2H), 1.62 (s, CH$_3$, 18H).

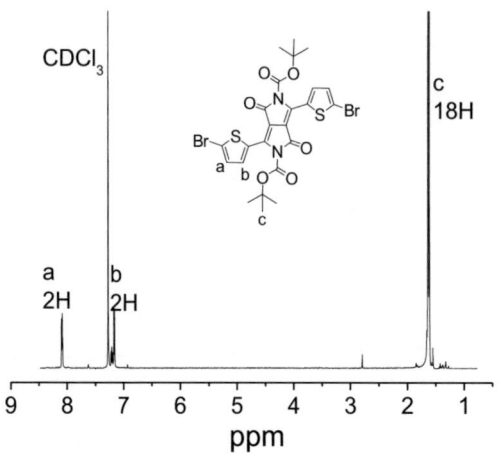

2,5-Bis(2-ethylhexyl)-3,6-di(thiophen-2-yl)pyrrolo[3,4-c]pyrrole-1,4(2H,5H)-dione (EH-ThDPP)[189]

3,6-Di(thiophen-2-yl)pyrrolo[3,4-c]pyrrole-1,4(2H,5H)-dione (**ThDPP**) (1.5 g, 4.99 mmol) and anhydrous potassium carbonate (2.07 g, 14.98 mmol) were dissolved in DMF (50 mL) in a two-neck round flask and heated to 145 °C under notrogen. 2-Ethylhexyl bromide (4.45 ml,

25 mmol) was injected one portion by syringe. The reaction mixture was stirred for 15 h at the same temperature. After cooling, the mixture was poured in 500 ml ice-water, and then filtered. The filter cake was washed by water and methanol several times. After drying in vacuum, the crude product was purified by silica gel chromatography using DCM/hexane as eluent to obtain a purple-black solid powder (1.84 g, 70%). ^{1}H-NMR (CDCl$_3$, 300 MHz): 8.91 (d, aromatic H, 2H), 7.65 (d, aromatic H, 2H), 7.28 (t, aromatic H, 2H), 4.06 (m, 4H), 1.88 (m, N-CH$_2$-H, 2H), 1.36−1.22 (m, CH$_2$-H, 16H), 0.89 (m, CH$_3$-H, 12H).

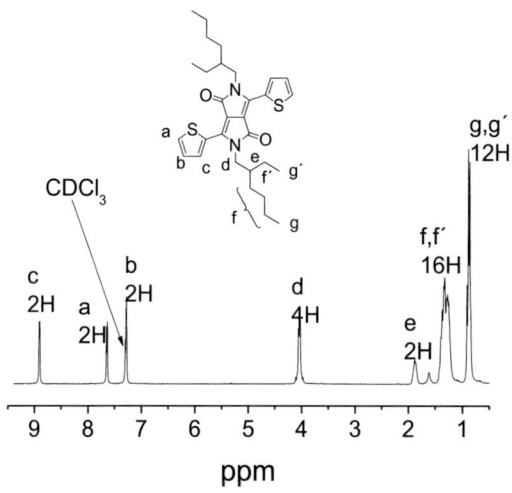

3,6-Bis(5-bromothiophen-2-yl)-2,5-bis(2-ethylhexyl)pyrrolo[3,4-c]pyrrole-1,4(2H,5H)-dione (Br-ET-ThDPP1)[189]

The same procedure as for **BrBoc-ThDPP** was used except that by **BrBoc-ThDPP** was replaced by **EH-ThDPP1**. A dark red solid was obtained (85%). ^{1}H-NMR (CDCl$_3$, 300 MHz): 8.67 (d, aromatic H, 2H), 7.24 (d, aromatic H, 2H), 3.97 (m, 4H), 1.87 (m, N-CH$_2$-H, 2H), 1.37−1.22 (m, CH$_2$-H, 16H), 0.90 (m, CH$_3$-H, 12H).

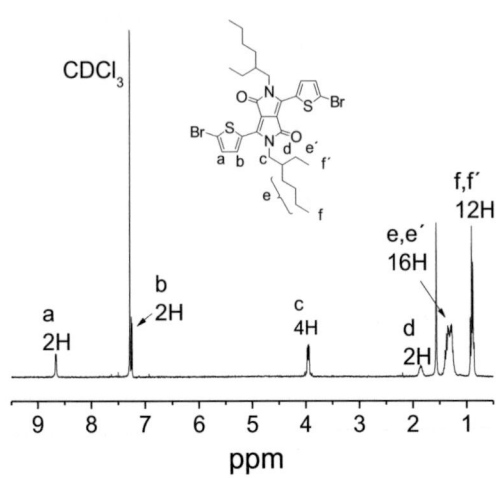

Poly(Boc-thienylDPP-thiophene) (P-Boc-ThDPP1)

In a Schlenk flask, di-tert-butyl-3,6-bis(5-bromothiophen-2-yl)-1,4-dioxopyrrolo[3,4-c]pyrrole-2,5(1H,4H)-dicarboxylate (**BrBoc-ThDPP**) (200 mg, 0.3 mmol), 2,5-bis(4,4,5,5-tetramethyl-1,3,2-dioxaborolan-2-yl)thiophene (98 mg, 0.3 mmol), and tetrakis(triphenylphosphine)-palladium(0) (10 mg, 9,1 μmol) were dissolved in 10 ml dry THF. The reaction mixture was degassed and heated to 50 °C under nitrogen. Then a solution of potassium carbonate (168 mg, 1.22 mmol) in 2 ml distillated water was added. The reaction mixture was refluxed under stirring for 2 h. After cooling, the reaction mixture was diluted with 50 ml DCM, and washed with 50 ml brine and 50 ml water twice. The organic phase was separated, and dried over MgSO$_4$, and the solvent was evaporated. The product was dissolved in a small amount DCM again and precipitated in methanol, yielding a dark solid (123 mg, 70%).%). UV/vis: λ_{max} = 541 nm. ^{1}H-NMR (300 MHz, CDCl$_3$): δ = 8.40 - 8.08 (m, aromatic H), 7.18-7.14 (m, aromatic H), 1.63 -1.58 (m, CH$_3$-H). Molecular weight: 21 000 Da. PD: 4.1.

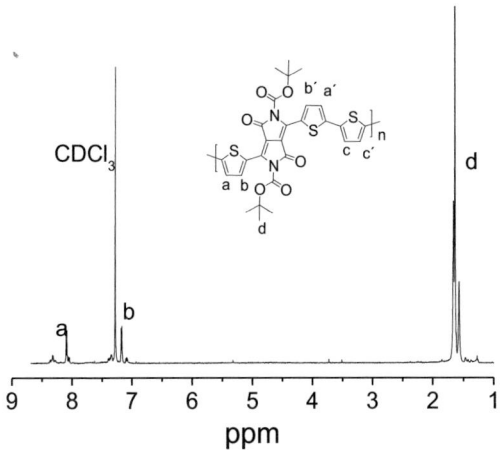

Poly(Boc-thienylDPP-ethylhexylthienylDPP-thiophene) (P-Boc-ThDPP2)

In a Schlenk flask, di-tert-butyl-3,6-bis(5-bromothiophen-2-yl)-1,4-dioxopyrrolo[3,4-c]pyrrole-2,5(1H,4H)-dicarboxylate (**BrBoc-ThDPP**) (100 mg, 0.15 mmol), 3,6-bis(5-bromothiophen-2-yl)-2,5-bis(2-ethylhexyl)pyrrolo[3,4-c]pyrrole-1,4(2H,5H)-dione (**Br-ET-ThDPP1**) (104, 0.15 mmol), 2,5-bis(4,4,5,5-tetramethyl-1,3,2-dioxaborolan-2-yl)thiophene 102 mg, 0.30 mmol), and tetrakis(triphenylphosphine)palladium(0) (5 mg, 4.6 μmol) were dissolved in 10 ml dry THF. The mixture was degassed and heated to 50 °C under nitrogen. Then a solution of potassium carbonate (168 mg, 1.22 mmol) in 2 ml distillated water was added. The reaction mixture was refluxed for 2h. After cooling, the reaction mixture was diluted with 50 ml DCM, and washed with 50 ml brine and 50 ml water twice. The organic phase was separated, and dried over MgSO$_4$, and the solvent was evaporated. The product was dissolved in a small amount DCM again and precipitated in methanol, yielding a dark solid (133 mg, 74%). UV/vis: λ_{max} = 612 nm. ^{1}H-NMR (300 MHz, CDCl$_3$): δ = 8.67 (m, aromatic H), 8.10 (m, aromatic H), 7.30-7.14 (m, aromatic H), 3.96 (m, N-CH$_2$-H), 2.19 (m, CH-H), 1.60 -1.20 (m, CH$_3$-H), 0.98 - 0.80 (m, CH$_3$-H). Molecular weight: 18 000 Da. PD: 4.2.

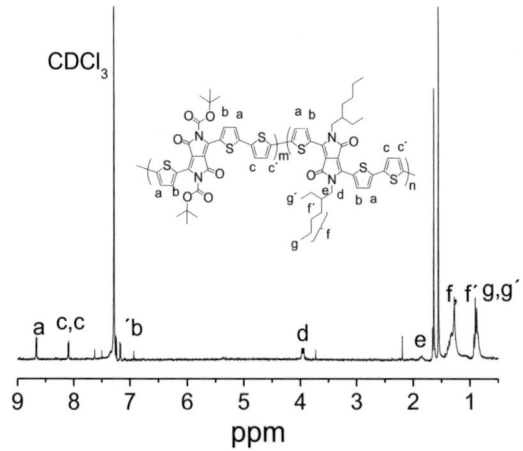

Poly(Boc-thienylDPP-ethylhexylthienylDPP-thiophene) (P-Boc-ThDPP3)

m:n = 2:1

The same procedure as for **P-Boc-ThDPP2** was used except that di-tert-butyl-3,6-bis(5-bromothiophen-2-yl)-1,4-dioxopyrrolo[3,4-c]pyrrole-2,5(1H,4H)-dicarboxylate (**BrBoc-ThDPP**) and 3,6-bis(5-bromothiophen-2-yl)-2,5-bis(2-ethylhexyl)pyrrolo[3,4-c]pyrrole-1,4(2H,5H)dione (**Br-ET-ThDPP1**) were 2:1 stoichiometrically. A dark solid was obtained (70%). UV/vis: λ_{max} = 600 nm. ^{1}H-NMR (300 MHz, CDCl$_3$): δ = 8.60 (m, aromatic H), 8.10 (m, aromatic H), 7.30-7.0 (m, aromatic H), 4.20 - 3.80 (m, N-CH$_2$-H), 2.19 (m, CH-H), 1.60 -1.0 (m, CH$_3$-H), 0.98 - 0.82 (m, CH$_3$-H). Molecular weight: 24 000 Da. PD: 4.4.

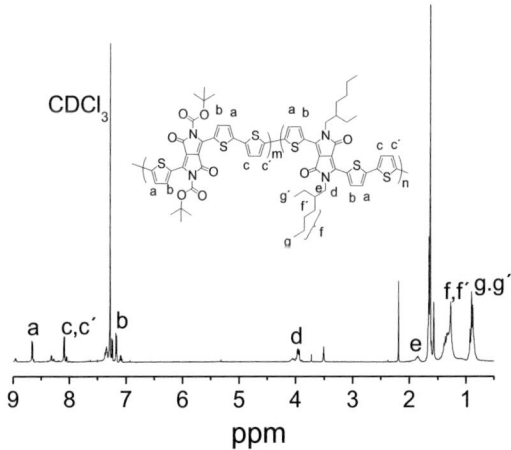

Poly(Boc-thienylDPP-ethylhexylthienylDPP-thiophene) (P-Boc-ThDPP4)

The same procedure as for **P-Boc-ThDPP2** was used except that di-tert-butyl-3,6-bis(5-bromothiophen-2-yl)-1,4-dioxopyrrolo[3,4-c]pyrrole-2,5(1H,4H)-dicarboxylate (**BrBoc-ThDPP**) and 3,6-bis(5-bromothiophen-2-yl)-2,5-bis(2-ethylhexyl)pyrrolo[3,4-c]pyrrole-1,4(2H,5H)dione (**Br-ET-ThDPP1**) were 2:1 stoichiometrically. A dark solid was obtained (70%). UV/vis: λ_{max} = 600 nm. ^{1}H-NMR (300 MHz, CDCl$_3$): δ = 8.66 (m, aromatic H), 8.09 (m, aromatic H), 7.30-7.10 (m, aromatic H), 3.90 - 3.80 (m, N-CH$_2$-H), 1.90 (m, CH-H), 1.60 -1.0 (m, CH$_3$-H), 0.98 - 0.80 (m, CH$_3$-H). Molecular weight: 18 000 Da. PD: 4.2.

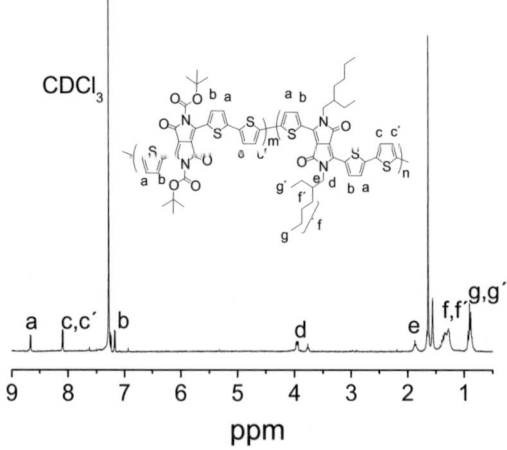

3,6-Di(thiophen-3-yl)pyrrolo[3,4-c]pyrrole-1,4(2H,5H)-dione (3-ThDPP)

The same procedure as for **ThDPP** was used except that 2-thiophenecarbonitrile was replaced by 3-thiophenecarbonitrile. A dark red solid was obtained (75%). The crude product was used directly for further reactions.

2-(2-Ethylheptyl)-5-(2-ethylhexyl)-3,6-di(thiophen-3-yl)pyrrolo[3,4-c]pyrrole-1,4(2H,5H)-dione

The same procedure was used as for 2,5-bis(2-ethylhexyl)-3,6-di(thiophen-2-yl)pyrrolo[3,4-c]pyrrole-1,4(2H,5H)-dione (**EH-ThDPP**) except that 3,6-di(thiophen-2-yl)pyrrolo[3,4-c]pyrrole-1,4(2H,5H)-dione (**ThDPP**) was replaced by 3,6-di(thiophen-3-yl)pyrrolo[3,4-c]pyrrole-1,4(2H,5H)-dione (**3-ThDPP**). A red solid was obtained (66%). ^{1}H-NMR (CDCl$_3$, 300 MHz): 8.74 (s, aromatic H, 2H), 8.50 (s, aromatic H, 1H), 8.16 (s, aromatic H, 1H), 7.94 (s, aromatic H, 1H), 7.79 (d, aromatic H, 2H), 3.91 (m, N-CH$_2$-H, 4H), 1.60 - 1.00 (m, alkyl-H, 18 H), 1.0 - 0.41 (m, CH$_3$-H, 12H).

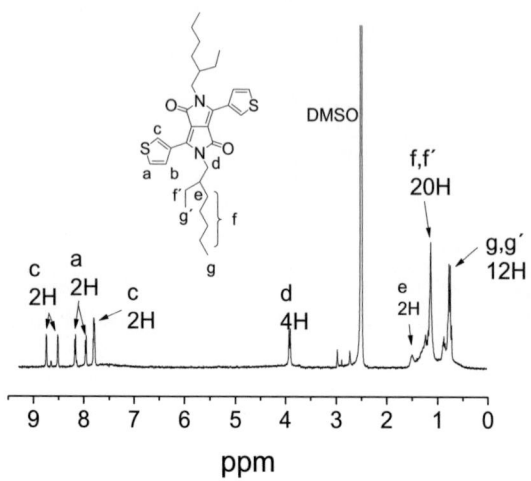

5.3.2 Conjugated polymers based on benzodifuranone

5.3.2.1 Conjugated polymers based on symmetrical benzodifuranone

3,7-Diphenylbenzo[1,2-b:4,5-b']difuran-2,6-dione (BZDF1)[66]

Using a Dean Stark apparatus, hydroquinone (1 g, 9.09 mmol) and mandelic acid (2.76 g 18.18 mmol) were dissolved in 1,2,4-trichlorobenzene (20 ml) and stirred for 12 h at 200 °C. After cooling to 60 °C, nitrobenzene (2.3 ml) was added and the mixture was stirred for another hour at 200 °C. After cooling to room temperature, 50 ml methanol were added. A precipitate formed, which was filtered off, washed with methanol and dried in air. The title compound was obtained as a yellow solid (2.32 g, 75%). ^{1}H-NMR (300 MHz, CDCl$_3$): δ = 7.80 (d, aromatic H, 4H), 7.55 (t, aromatic H, 4H), 6.89 (t, aromatic H, 2H), 6.61 (s, aromatic H, 2H). UV/Vis (DCM): 469 nm. Fluorescence (DCM): 600 nm. ε (469) = 35 000 L mol^{-1}cm^{-1}.

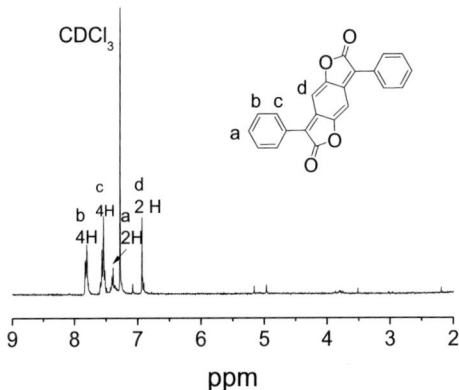

3,7-Bis(4-bromophenyl)benzo[1,2-b:4,5-b']difuran-2,6-dione (BZDF2)

The same procedure as described for compound **BZDF1** was used except that mandelic acid was replaced by 4-bromo-mandelic acid. **BZDF2** was obtained as a dark greenish solid (65%), which was only little soluble in common solvents at room temperature. (m.p. above 300 °C). Anal. Calculated for C$_{22}$H$_{10}$Br$_2$O$_4$: C, 53.05%; H, 2.02%. Found: C, 52.80%; H, 3.08%. UV/Vis (DCM): λ$_{max}$ at 483 nm. Fluorescence (DCM): 586 nm. Mp > 300 °C.

3,7-Bis(5-bromo-2-methoxyphenyl)benzo[1,2-b:4,5-b']difuran-2,6-dione (BZDF3)

The same procedure as described for compound **BZDF2** was used except that mandelic acid was replaced by 3-bromo-5-methoxy-mandelic acid. **BZDF3** was obtained as a dark greenish solid (65%), which was only little soluble in common solvents at room temperature. (m.p. above 300 °C). Anal. Calculated for $C_{24}H_{14}Br_2O_6$: C, 51.64%; H, 2.53%. Found: C, 51.10%; H, 3.01%. UV/Vis (DCM): λ_{max} at 391, 477 nm. Fluorescence (DCM): 610 nm. Mp > 300 °C.

3,7-Bis(3-bromo-4,5-dimethoxyphenyl)benzo[1,2-b:4,5-b']difuran-2,6-dione (BZDF4)

The same procedure as described for compound **BZDF1** was used except that mandelic acid was replaced by 3-bromo-4,5-dimethoxy-mandelic acid. **BZDF4** was obtained as a dark solid (70%), which was only little soluble in common solvents at room temperature. (m.p. above 300 °C). Anal. Calculated for $C_{26}H_{18}Br_2O_8$: C, 50.51%; H, 2.93%. Found: C, 50.00%; H, 3.52%. UV/Vis (DCM): λ_{max} at 382, 521 nm. Fluorescence (DCM): 606 nm. Mp > 300 °C.

3,7-Bis(4-(2,3-dihydrothieno[3,4-b][1,4]dioxin-5-yl)phenyl)benzo[1,2-b:4,5-b']difuran-2,6-dione (EDOT-BZDF)

In a vial, 200 mg (0.40 mmol) 3,7-bis(4-bromophenyl)benzo[1,2-b:4,5-b']difuran-2,6-dione (**BZDF2**), 305 mg (1.00 mmol) 3,4-ethylenedioxythien-2-yl trimethylstannane and 14 mg (0.012 mmol) tetrakis(triphenylphosphine) palladium(0) were dissolved in 5 ml dry DMF and stirred for 5 min. The mixture was degassed, and heated in the microwave synthesizer at 160 °C for 1 h. After cooling the mixture was diluted with 50 ml DCM, and washed with 50 ml brine and 50 ml water. The organic layer was separated, dried over magnesium sulfate and evaporated. The dark product was recrystallized from DCM/methanol. Yield: 196 mg (79%) ^{1}H-NMR (400 MHz, CDCl$_3$): δ = 7.87 (d, aromatic H, 8H), 6.96 (s, EDOT aromatic H, 2H), 6.41 (s, aromatic H, 1H), 4.33 (t, EDOT-CH$_2$, 8H). UV/Vis (DCM): 298, 391, 590 nm. ε(590) = 63 580 L mol^{-1}cm^{-1}.

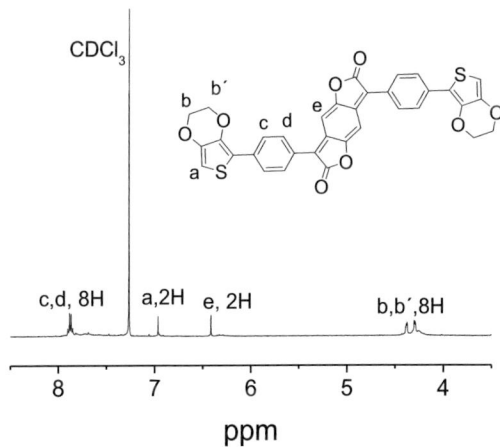

3,7-Bis(4-(2,3-dihydrothieno[3,4-b][1,4]dithien-5-yl)phenyl)benzo[1,2-b:4,5-b']difuran-2,6-dione (EDTT-BZDF)

The same procedure as described for compound **EDOT-BZDF** was used except that 3,4-ethylenedioxythien-2-yl trimethylstannane was replaced by 3,4-ethylenedithiathien-2-yl trimethylstannane. A dark solid (200 mg, 89%) was obtained. ^{1}H-NMR (400 MHz, CDCl$_3$): δ = 7.85 (d, aromatic H, 8H), 6.69 (s, EDTT aromatic H, 2H), 7.09 (s, aromatic H, 1H), 3.24 (t, EDTT-CH$_2$, 8H). UV/Vis (DCM): 306, 502 nm. ε(502) = 79 220 L mol^{-1}cm^{-1}.

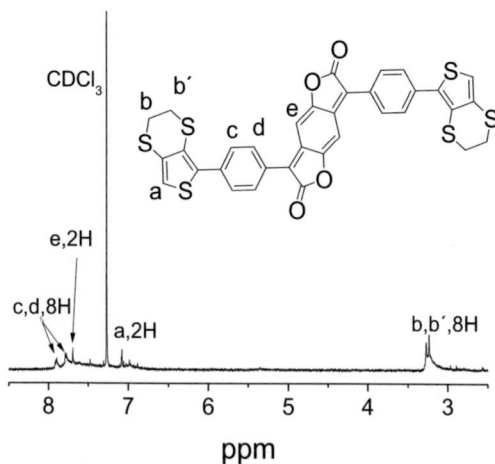

3,6-Diphenylbenzo[1,2-b:6,5-b']difuran-2,7-dione (BZDF5)[66]

Using a Dean Stark apparatus, *ortho*-quinone (1.20 g, 10.82 mmol) and mandelic acid (5.0 g 21.64 mmol) were dissolved in 1,2,4-trichlorobenzene (30 ml) and stirred for 5 h at 200 °C. After cooling to 60 °C, nitrobenzene (5.0 ml) was added and the mixture was stirred for another hour at 200 °C. After cooling to room temperature, 50 ml methanol were added. The solid formed was filtered off, washed with methanol and dried in air. **BZDF5** was obtained as a dark red solid (3.28 g, 61%). ^{1}H-NMR (400 MHz, CDCl$_3$): δ = 7.82 (d, aromatic H, 4H), 7.49 (q, aromatic H, 4H), 7.04 (t, aromatic H, 2H), 6.92 (d, aromatic H, 2H).

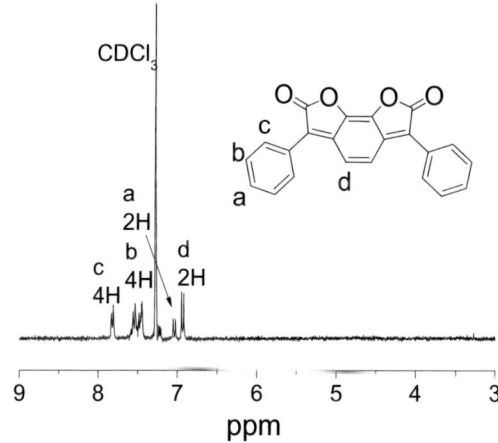

3,6-Bis(4-bromophenyl)benzo[1,2-b:6,5-b']difuran-2,7-dione (BZDF6)

Using a Dean Stark apparatus, *ortho*-quinone (1.20 g, 10.82 mmol) and 4-bromo-mandelic acid (5.0 g 21.64 mmol) were dissolved in 1,2,4-trichlorobenzene (30 ml) and stirred for 5 h at 200°C. After cooling to 60 °C, nitrobenzene (5.0 ml) was added and the mixture was stirred for another hour at 200°C. After cooling to room temperature, 50 ml methanol were added. The solid formed was filtered off, washed with methanol and dried in air. **BZDF6** was obtained as a dark red solid (3.28 g, 61%). m.p.: over 300 °C. Anal. Calculated for $C_{22}H_{10}Br_2O_4$: C, 53.05%; H, 2.02%. Found: C, 52.10%; H, 2.90%. UV/Vis (DCM): λ_{max} at 493 nm. Mp > 300 °C.

3,6-Bis(4-(2,3-dihydrothieno[3,4-b][1,4]dioxin-5-yl)phenyl)benzo[1,2-b:6,5-b']difuran-2,7-dione (EDOT-oBZDF)

The same procedure as described for compound **EDOT-BZDF** was used, except that 3,7-bis(4-bromophenyl)benzo[1,2-b:4,5-b']difuran-2,6-dione (**BZDF2**) was replaced by 3,6-bis(4-bromophenyl)benzo[1,2-b:6,5-b']difuran-2,7-dione (**BZDF6**). A dark solid (93 mg, 82%) was obtained. ^{1}H-NMR (400 MHz, CDCl$_3$): δ = 7.88 (d, aromatic H, 8H), 6.98 (s, EDOT aromatic H H, 2H), 7.29 (s, aromatic H, 1H), 4.34 (t, EDOT-CH$_2$, 8H). UV/Vis (DCM): 302, 588 nm. ε(588) = 23 080 L mol^{-1}cm^{-1}.

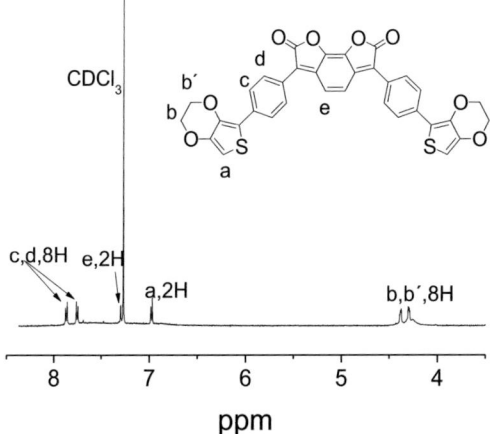

3,6-Bis(4-(2,3-dihydrothieno[3,4-b][1,4]dithien-5-yl)phenyl)benzo[1,2-b:6,5-b']difuran-2,7-dione (EDTT-oBZDF)

The same procedure as described for compound **EDOT-BZDF** was followed except that 3,4-ethylenedioxythien-2-yl trimethylstannane was replaced by 3,4-ethylenedithiathien-2-yl trimethylstannane. A dark solid (120 mg, 86%) was obtained. ^{1}H-NMR (400 MHz, CDCl$_3$): δ = 7.84 (d, aromatic H, 8H), 7.44 (s, EDTT aromatic H, 2H), 7.09 (s, aromatic H, 1H), 3.27 (t, EDTT-CH$_2$, 8H). UV/Vis (DCM): 307, 407, 558 nm. ε(558) = 115 560 L mol^{-1}cm^{-1}.

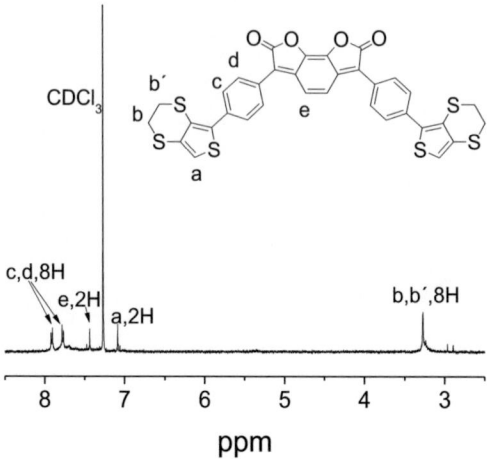

3,7-Bis(4-(4-octylthiophen-2-yl)phenyl)benzo[1,2-b:4,5-b']difuran-2,6-dione (BZDF7)

The same procedure as described for compound **EDOT-BZDF** was used except that 3,4-ethylenedioxythien-2-yl trimethylstannane was replaced by trimethyl(4-octylthiophen-2-yl)-stannane. A dark solid (260 mg, 90%) was obtained. ^{1}H-NMR (300 MHz, CDCl$_3$): δ = 7.88 (d, aromatic H, 8H), 7.30 (d, aromatic H, 4H), 6.98 (s, thiopehne aromatic H, 2H), 2.66 (t, CH$_2$-H, 4H), 1.76 - 1.33, (m, akyl H, 28 H) 0.90 (t, CH$_3$-H, 6H). UV/Vis (DCM): 320, 390, 550 nm. ε(550) = 56 300 L mol^{-1}cm^{-1}.

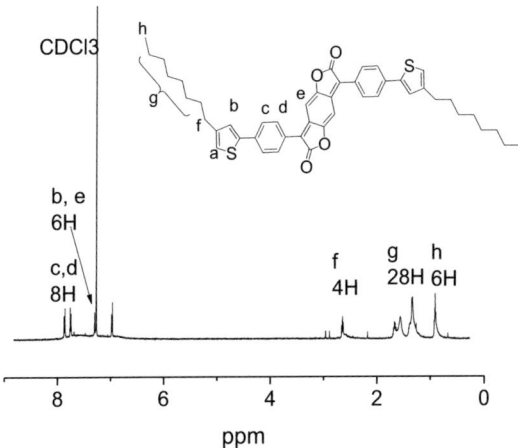

3,7-Bis(4-(5-bromo-4-octylthiophen-2-yl)phenyl)benzo[1,2-b:4,5-b']difuran-2,6-dione (BZDF9)

Under nitrogen, a flask was charged with 3,7-Bis(4-(4-octylthiophen-2-yl)phenyl)benzo[1,2-b:4,5-b']difuran-2,6-dione (**BZDF7**) (200 mg, 0.27 mmol) in 50 mg DCM. NBS (122 mg, 0.69 mmol) was added in one pot. The reaction mixture was stirred at room temperature over night. The solution was then washed with 50 ml water twice, and the organic phase was dried over MgSO$_4$. After evaporating the solvent, the crude product was recystallized from DCM/methanol. A dark solid was obtained (195 mg, 80%). ^{1}H-NMR (300 MHz, CDCl$_3$): δ = 7.83 (d, aromatic H, 8H), 7.35 (d, aromatic H, 4H), 2.67 (t, CH$_2$, 4H), 1.80 - 1.30, (m, akyl H, 28 H) 0.95 (t, CH$_3$-H, 6H). UV/Vis (DCM): 556 nm. ε(556) = 59 450 L mol^{-1}cm^{-1}.

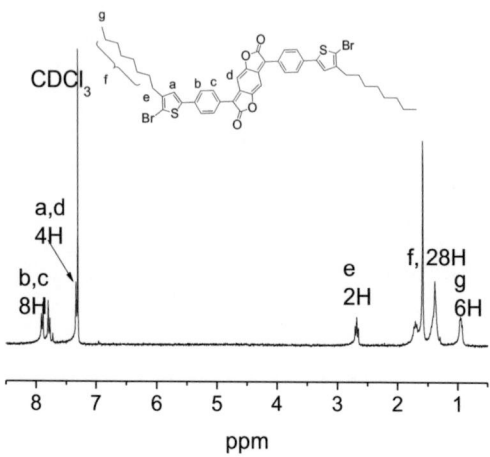

3,6-Bis(4-(4-octylthiophen-2-yl)phenyl)benzo[1,2-b:6,5-b']difuran-2,7-dione (BZDF8)

The same procedure as described for compound **EDOT-oBZDF** was used except that 3,4-ethylenedioxythien-2-yl trimethylstannane was replaced by trimethyl(4-octylthiophen-2-yl)-stannane. A dark solid (160 mg, 84%) was obtained. ^{1}H-NMR (300 MHz, CDCl$_3$): δ = 7.89 (d, aromatic H, 4H), 7.81 (d, aromatic H, 4H), 7.33 (s, aromatic H, 2H), 7.00 (s, aromatic H, 2H), 6.57 (d, aromatic H, 2H), 2.66 (t, CH$_2$, 4H), 1.76 - 1.33, (m, akyl H, 24 H) 0.90 (t, CH$_3$-H, 6H). UV/Vis (DCM): 551 nm. ε(551) = 50 060 L mol^{-1}cm^{-1}.

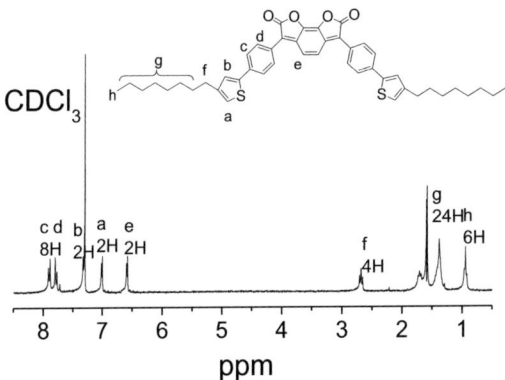

3,6-bis(4-(5-bromo-4-octylthiophen-2-yl)phenyl)benzo[1,2-b:6,5-b']difuran-2,7-dione (BZDF10)

The same procedure as described for compound **BZDF9** was used except that 3,7-Bis(4-(4-octylthiophen-2-yl)phenyl)benzo[1,2-b:4,5-b']difuran-2,6-dione (**BZDF7**) was replaced by 3,6-Bis(4-(4-octylthiophen-2-yl)phenyl)benzo[1,2-b:6,5-b']difuran-2,7-dione (**BZDF8**). A dark solid was obtained (137 mg, 75%). ^{1}H-NMR (300 MHz, CDCl$_3$): δ = 7.88 (d, aromatic H, 8H), 7.30 (d, aromatic H, 4H), 6.98 (s, thiopehne aromatic H, 2H), 2.66 (t, CH$_2$, 4H), 1.76 - 1.33, (m, akyl H, 28 H) 0.90 (t, CH$_3$-H, 6H). UV/Vis (DCM): 553 nm. ε(553) = 51 670 L mol^{-1}cm^{-1}.

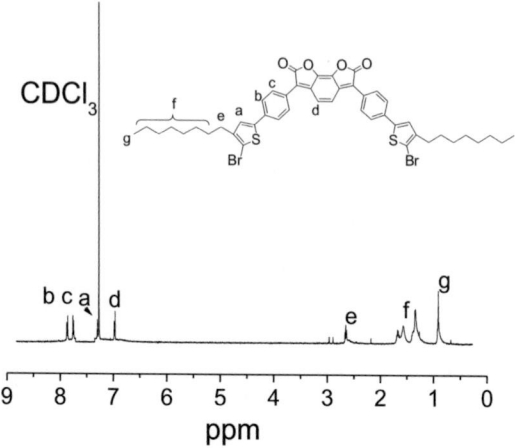

Poly(benzodifuranone-3-octylthiophene) (P-BZDF-TH1)

In a Schlenk flask, 3,7-bis(4-(5-bromo-4-octylthiophen-2-yl)phenyl)benzo[1,2-b:4,5-b']difuran-2,6-dione (**BZDF9**) (100 mg, 0.11 mmol), 2,5-bis(4,4,5,5-tetramethyl-1,3,2-dioxaborolan-2-yl)thiophene (38 mg, 0.11 mmol), $Pd_2(dba)_3$ (3 mg, 3,7 μmol) tri-*t*-butyl phosphine (1.4 mg, 6.8 mmol) were dissolved in 10 ml dry THF. The solution was degassed and heated to 50 °C under stirring. A degassed solution of K_2CO_3 (62 mg, 0.45 mmol) in 2 ml water was added to the reaction mixture. The mixture was stirred at 70 °C for 3 h. After cooling, the dark solution was diluted with 50 ml DCM, and washed with 40 ml saturated brine and 50 ml water. The organic phase was dried over $MgSO_4$, and filtered through celite. The solvent was evaporated. The product was precipitated in methanol, yielding 65 mg dark solid (71%). ^{1}H-

NMR (300 MHz, CDCl$_3$): δ = 7.80 - 7.20 (m, aromatic H), 2.70 (m, α-CH$_2$-H), 1.76 - 1.33, (m, CH$_2$-H) 0.90 (m, CH$_3$-H). UV/Vis (DCM): 578 nm. ε(578) = 36 600 L mol^{-1} cm^{-1}. Molecular weight: 12 000 Da. Polydispersity: 1.6.

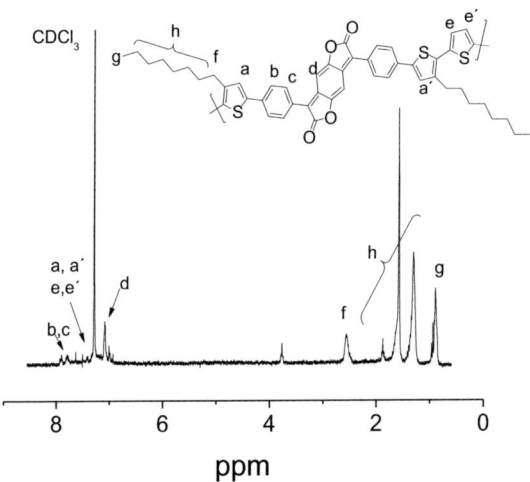

Poly(ortho-benzodifuranone-3-octylthiophene) (P-oBZDF-TH1)

The same procedure as described for polymer **P-BZDF-TH** was used except that 3,7-bis(4-(5-bromo-4-octylthiophen-2-yl)phenyl)benzo[1,2-b:4,5-b']difuran-2,6-dione (**BZDF9**) was replaced by 3,6-bis(4-(5-bromo-4-octylthiophen-2-yl)phenyl)benzo[1,2-b:6,5-b']difuran-2,7-dione (**BZDF10**). A dark solid was obtained (60%). ^{1}H-NMR (300 MHz, CDCl$_3$): δ = 7.96 - 6.82 (m, aromatic H), 2.56 (m, α-CH$_2$-H), 1.75 - 1.04, (m, CH$_2$-H) 0.89 (m, CH$_3$-H). UV/vis

(DCM): 563 nm. ε(563) = 31 500 L mol^{-1} cm^{-1}. Molecular weight: 10 500 Da. Polydispersity: 2.1.

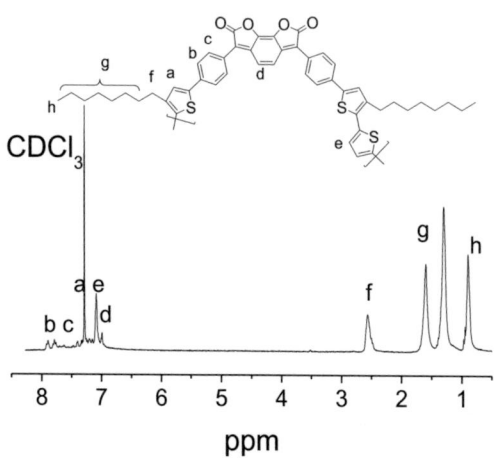

Poly(benzodifuranone-3-octylthiophene) (P-BZDF-TH2)

Under N$_2$, FeCl$_3$ (118 mg, 0.73 mmol) was dissolved in dry DCM (20 ml). A solution of **3,7-bis(4-(4-octylthiophen-2-yl)phenyl)benzo[1,2-b:4,5-b']difuran-2,6-dione** (**BZDF7**) (200 mg, 0.27 mmol) in 5 ml dry DCM was added. The reaction mixture was stirred for a further hour at room temperature. The mixture was diluted with 50 ml DCM and 20 ml 1 M HCl were added. The organic phased was washed twice with brine and water and then dried. The solvent was removed and the crude product was precipitated in methanol, and washed with methanol in a Soxhlet over night. After drying, a dark product was obtained (142 mg, 70%). ^{1}H-NMR (300 MHz, CDCl$_3$): δ = 7.90 - 6.90 (m, aromatic H), 2.60 (m, α-CH$_2$-H), 1.76 - 1.10, (m, CH$_2$-H)

0.90 (m, CH$_3$-H). UV/vis (DCM): λ_{max} = 575 nm. Molecular weight: 16 000 Da, Polydispersity: 1.8.

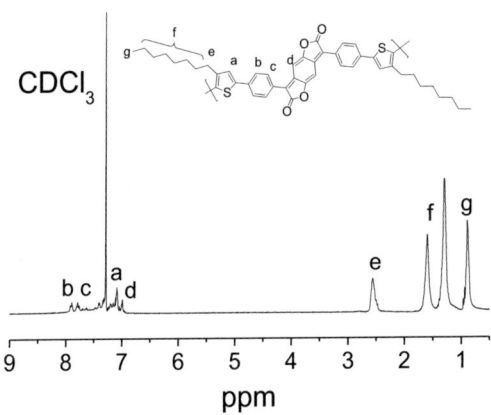

Poly(*ortho*benzodifuranone-3-octylthiophene) (P-oBZDF-TH2)

The same procedure as for P-BZDF-TH2 was used except that **BZDF7** was replaced by **BZDF8**. A dark solid was obtained yielding 65%. ^{1}H-NMR (300 MHz, CDCl$_3$): δ = 7.86 - 6.80 (m, aromatic H), 2.60 (m, α-CH$_2$-H), 1.76 - 1.10, (m, CH$_2$-H) 0.90 (m, CH$_3$-H). UV/vis (DCM): λ_{max} = 578 nm. Molecular weight: 11 000 Da, Polydispersity: 2.2.

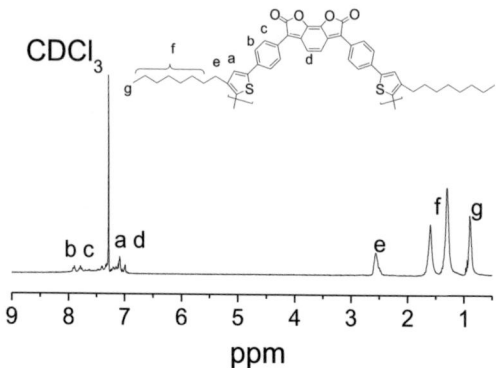

5.3.2.2 Conjugated polymers based on unsymmetrical benzodifuranone

5-Hydroxy-3-phenylbenzofuran-2(3H)-one[66]

In a flask, hydroquinone (7.24 g, 66 mmol) and mandelic acid (5 g, 32 mmol) were dissolved in 20 ml 73% sulfuric acid. The mixture was heated to 120 °C and stirred for two hours at the same temperature. After cooling to room temperature, the reaction mixture was poured in 100 ml ice water. A white solid precipitated. The solid was filtered off, and washed with cold water until it was acid-free. The crude product was dried under vacuum yielding 7 g (95%) of the title compound. ^{1}H-NMR (300 MHz, CDCl$_3$): δ = 7.37 (m. aromatic H, 3H), 7.24, (d, aromatic H, 2H), 7.08 (d, aromatic H, 2H), 6.83 (d, aromatic H, 2H), 6.72 (s, aromatic H, 1H), 4.88 (s, CH-H, 1H).

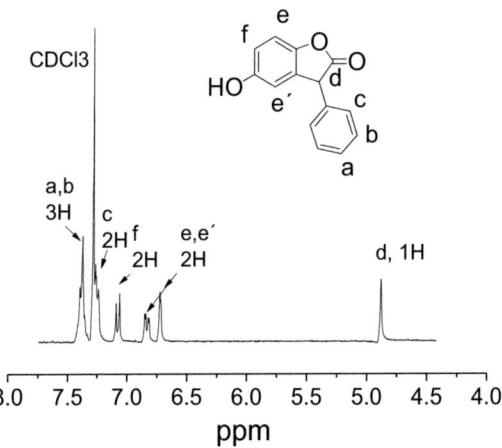

3-(4-Bromophenyl)-5-hydroxybenzofuran-2(3H)-one

The same procedure as described for 5-hydroxy-3-phenylbenzofuran-2(3H)-one was used except that mandelic acid was replaced by 4-bromo-mandelic acid. A white solid was obtained (90%). ^{1}H-NMR (300 MHz, CDCl$_3$): δ = 7.53 (d, aromatic H, 2H), 7.16 - 7.07 ((m, aromatic H, 3H), 6.85 (d, aromatic H, 1H), 6.72 (s, aromatic H, 1H), 4.84 (s, CH-H, 1H).

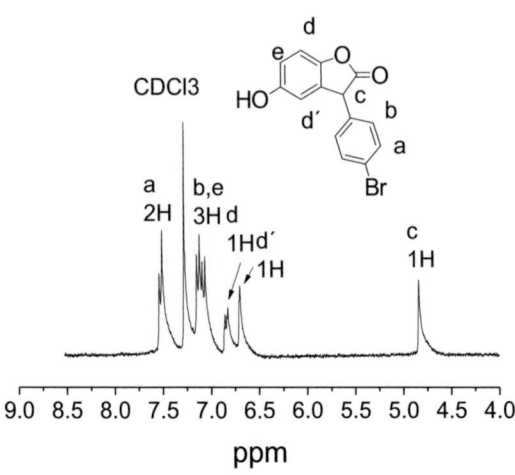

2-(3,4-Dimethoxyphenyl)-2-hydroxyacetic acid[190]

In a three neck flask, 3,4-dimethoxybenzaldehyde (2 g, 12.04 mmol) and tetraammonium bromide (0.19 g, 0.60 mmol) were dissolved in 1.45 ml chloroform under N_2. The mixture was heated to 50 °C under stirring. A solution of NaOH (2.41 g, 60.20 mmol) in 10 ml water was added dropwise. The reaction mixture was stirred at the same temperature for 15 h. After cooling, the mixture was diluted with 50 ml DCM. A solution of 1 M HCl was added until a pH-value of 3 was reached. The organic phase was dried over $MgSO_4$, and the solvent was removed. The crude product was purified using column chromatography (DCM/hexane:1/4). The title compound was obtained as a slightly yellow solid. (1.80 g, 74%). ^{1}H-NMR (300 MHz, $CDCl_3$): δ = 10.20 (s, OH-H, 1H), 7.73 (d, aromatic H, 1H), 7.50 (s, aromatic H, 1H), 6.51 (d, aromatic H, 1H), 5.30 (t, CH-H, 1H), 3.91 (d, CH_3-H, 6H).

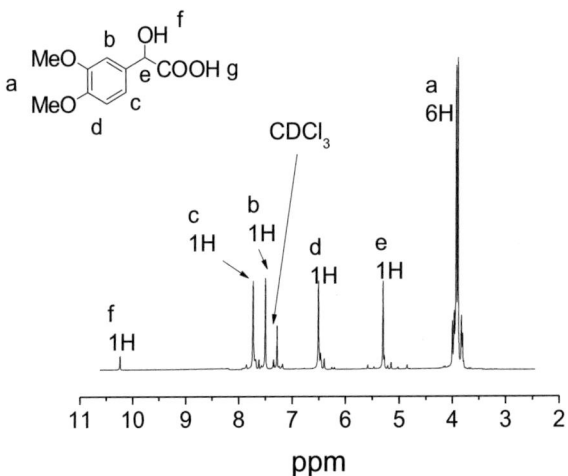

3-Bromo-4,5-dimethoxybenzaldehyde[191]

In a flask, 3,4-dimethoxybenzaldehyde (5 g, 30 mmol) was dissolved in 50 ml acetonitrile. The solution was cooled to 0 °C in an ice-water bath under N_2. NBS (5.89 g, 33 mmol) was added in small portions within a time period of 30 min. The reaction mixture was stirred over night whilst slowly warming to room temperature. The reaction mixture was poured in 20 ml water. A white solid was precipitated. The crude product was dried in air. A pure product was obtained using recrystallization from methanol/DCM. (6.78 g, 92%). ^{1}H-NMR (300 MHz, CDCl$_3$): δ = 9.94 (s, CHO-H, 1H), 7.04 (s, aromatic H, 1H), 6.72 (s, aromatic H, 1H), 3.87 (s, CH3-H, 6H).

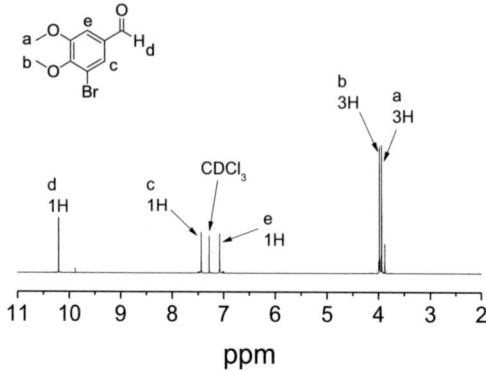

2-(3-Bromo-4,5-dimethoxyphenyl)-2-hydroxyacetic acid

The same procedure was used as for 2-(3,4-dimethoxyphenyl)-2-hydroxyacetic acid except that 3,4-dimethoxybenzaldehyde was replaced by 3-bromo-4,5-dimethoxybenzaldehyde. A slightly yellow solid was obtained. (70%). ^{1}H-NMR (300 MHz, CDCl$_3$): δ = 9.68 (s, OH-H, 1H), 8.05 (s, aromatic H, 1H), 7.52 (s, aromatic H, 1H), 5.96 (broad signal, 1H), 3.89 (d, CH$_3$-H, 6H). Mp = 136 °C.

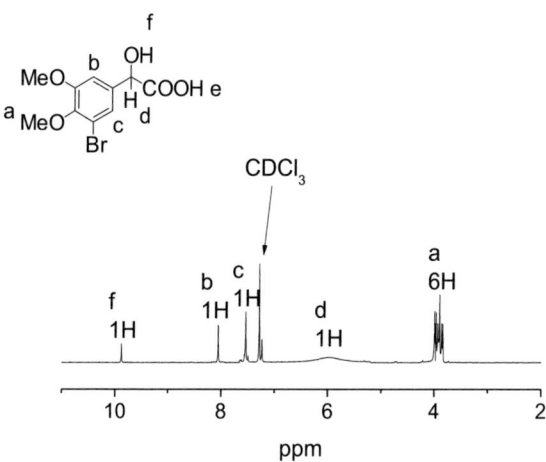

5-Bromo-2-methoxybenzaldehyde[191]

The same procedure was used as for 3-bromo-4,5-dimethoxybenzaldehyde except that 3,4-dimethoxybenzaldehyde was replaced by 2-methoxybenzaldehyde. A white solid was obtained. (96%). ^{1}H-NMR (300 MHz, CDCl$_3$): δ = 10.41 (s, CHO-H, 1H), 7.95 (s, aromatic H, 1H), 7.63 (d, aromatic H, 1H), 6.91 (d, aromatic H, 1H), 3.94 (s, CH$_3$-H, 3H).

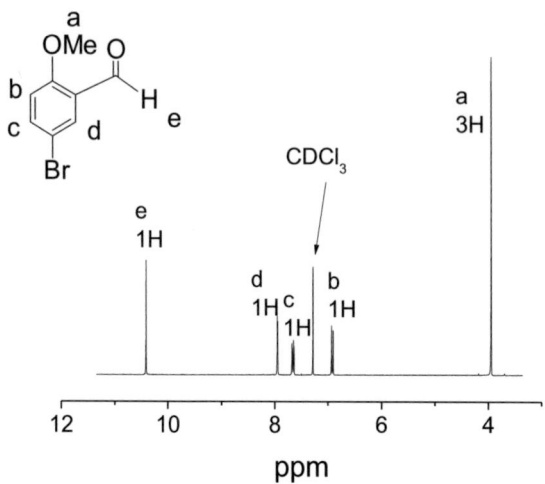

3-(3,4-Dimethoxyphenyl)-7-phenylbenzo[1,2-b:4,5-b']difuran-2,6-dione (BZDF15)

4,5-Dimethoxy-mandelic acid (1.20 g, 5.66 mmol), 5-hydroxy-3-phenylbenzofuran-2(3H)-one (1.27 g, 5.66 mmol), and p-toluenesulphonric acid (1.27 g, 6.79 mmol) were dissolved in 1,2,4-trichlorobenzene (20 ml), and heated to 80 °C for 6 h under stirring. The mixture was allowed to cool to 40 °C, and nitrobenzene (2 ml) was added. The temperature was raised to 100 °C and stirred for a further hour. Methanol (5 ml) was added. The precipitated product was filtered off and recrystallized from DCM/methanol. A dark red solid was obtained (1.47 g, 65%). ^{1}H-NMR (300 MHz, CDCl$_3$): δ = 7.81 (t, aromatic H, 4H), 7.55-7.44 (m, aromatic H, 5H), 7.06 (s, aromatic H, 1H), 6.93 (d, aromatic H H, 2H), 4.00 (s, CH$_3$-H, 6H). UV/Vis (DCM): 525 nm. Fluorescence (DCM): 635 nm. ε(525) = 42 000 L mol^{-1}cm^{-1}. Mp > 300 °C.

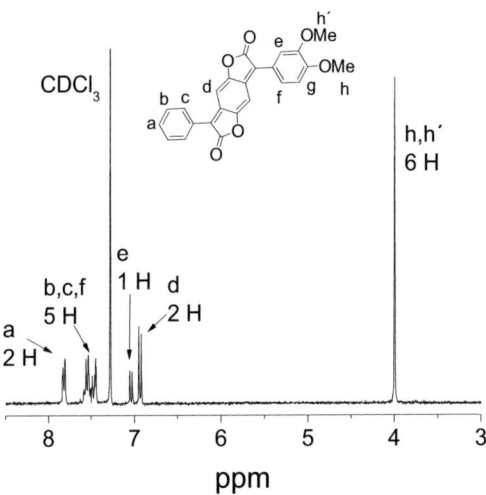

3-(3-Bromo-4,5-dimethoxyphenyl)-7-(4-bromophenyl)benzo[1,2-b:4,5-b']difuran-2,6-dione (BZDF16)

The same procedure as described for compound **BZDF15** was used except that 4,5-dimethoxy-mandelic acid and 5-hydroxy-3-phenylbenzofuran-2(3H)-one were replaced by 3-bromo-4,5-dimethoxy-mandelic acid and 3-(4-bromophenyl)-5-hydroxybenzofuran-2(3H)-one. **BZDF16** was obtained as a dark red solid (60%). ^{1}H-NMR (300 MHz, CDCl$_3$): δ = 7.70 (d, aromatic H, 4H), 7.19 (s, aromatic H H, 1H), 6.87 (d, aromatic H H, 2H), 6.61 (d, aromatic H H, 1H), 3.94 (d, CH$_3$-H, 6H). UV/Vis (DCM): 441, 490 nm. Fluorescence (DCM): 645 nm. Mp > 300 °C.

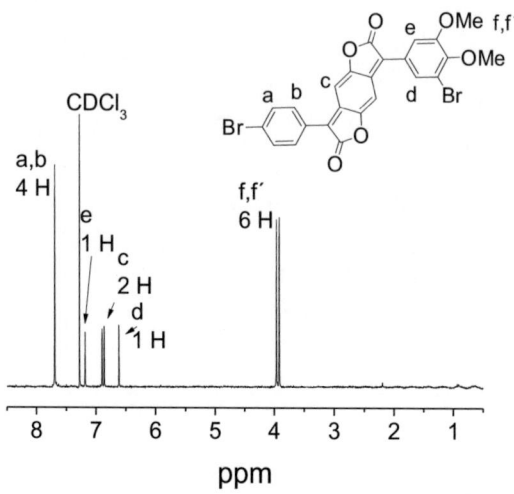

3-(5-Bromo-2-methoxyphenyl)-7-(4-bromophenyl)benzo[1,2-b:4,5-b']difuran-2,6-dione

The same procedure as described for compound **BZDF16** was used except that 3-bromo-4,5-dimethoxy-mandelic acid was replaced by 2-(3-bromo-5-methoxyphenyl)-2-hydroxyacetic acid. The title compound was obtained as a dark brown solid (50%). ^{1}H-NMR (300 MHz, CDCl$_3$): δ = 7.69 (d, aromatic H, 4H), 7.58 (d, aromatic H, 1H), 6.93 (d, aromatic H H, 1H), 6.84 (s, aromatic H H, 2H), 6.61 (s, aromatic H H, 1H), 3.90 (s, CH$_3$-H, 3H). UV/Vis (DCM): 476 nm. Mp > 300 °C.

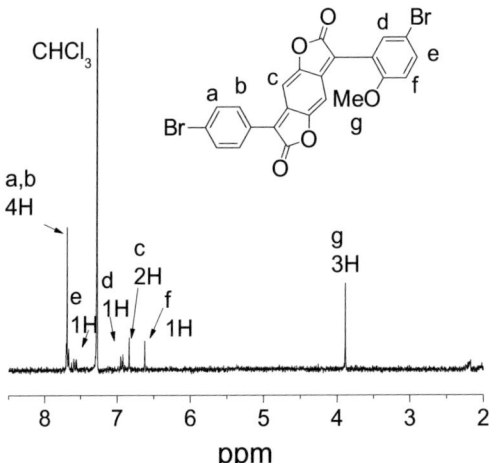

3-(5-Bromo-2-methoxyphenyl)-7-phenylbenzo[1,2-b:4,5-b']difuran-2,6-dione

The same procedure as described for 3-(5-bromo-2-methoxyphenyl)-7-(4-bromophenyl)-benzo[1,2-b:4,5-b']difuran-2,6-dione was used except that 3-(4-bromophonyl)-5-hydroxy-benzofuran-2(3H)-one was replaced by 5-hydroxy-3-phenylbenzofuran-2(3H)-one. The title compound was obtained as a dark brown solid (45%). ^{1}H-NMR (300 MHz, CDCl$_3$): δ = 7.80 (d, aromatic H, 2H), .68 (d, aromatic H, H), 7.55 (m, aromatic H, 3H), 6.91 (m, aromatic H H, 3H), 6.84 (s, aromatic H H, 2H), 3.89 (s, CH$_3$-H, 3H). UV/Vis (DCM): 466 nm. Mp > 300 °C.

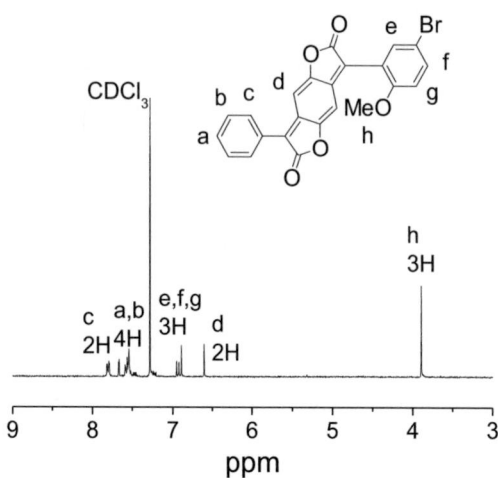

3-(3-(2,3-Dihydrothieno[3,4-b][1,4]dioxin-5-yl)-4,5-dimethoxyphenyl)-7-(4-(2,3-dihydro-thieno-[3,4-b][1,4]dioxin-5-yl)phenyl)benzo[1,2-b:4,5-b']difuran-2,6-dione (BZDF17)

3-(3-Bromo-4,5-dimethoxyphenyl)-7-(4-bromophenyl)benzo[1,2-b:4,5-b']difuran-2,6-dione (**BZDF16**) (22 mg, 0.38 mmol), (2,3-dihydrothieno[3,4-b][1,4]dioxin-5-yl)trimethylstannane (273 mg, 0.90 mmol), and Pd(PPh$_3$)$_4$ (20 mg, 0.02 mmol) were dissolved in 10 ml dry DMF. Under N$_2$, the solution was degassed and stirred at 150 °C for 2 h. After cooling, the reaction mixture was diluted with 50 ml DCM, and washed twice with 50 ml saturated brine and water. The organic phase was dried over MgSO$_4$, and the solvent was evaporated. The product was recrystallized from methanol/DCM yielding a dark solid. (240 mg, 80%). ^{1}H-NMR (300 MHz, CDCl$_3$): δ = 7.88 (d, aromatic H, 4H), 7.07 (s, aromatic H, 1H), 6.88 (s, aromatic H, 1H), 6.44

(s, aromatic H, 2H), 6.28 (s, thiophene-H, 2H), 4.35 (t, CH$_2$-H, 8H), 4.00 (s, CH$_3$-H, 6H). UV/vis (DCM): λ_{max} = 532, 323 nm.

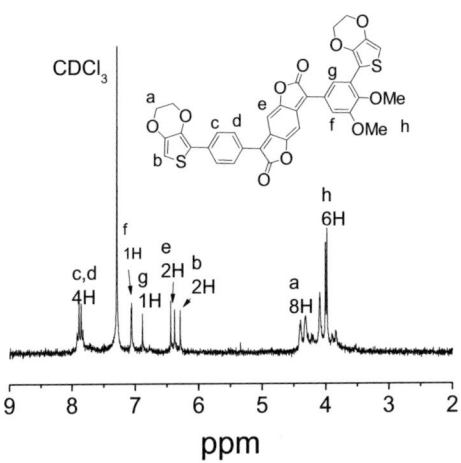

3-(3-([2,2'-Bithiophen]-5-yl)-4,5-dimethoxyphenyl)-7-(4-([2,2'-bithiophen]-5-yl)phenyl)-benzo[1,2-b:4,5-b']difuran-2,6-dione (BZDF18)

The same procedure was used as for **BZDF17** except that (2,3-dihydrothieno[3,4-b][1,4]dioxin-5-yl)trimethylstannane was replaced by [2,2'-bithiophen]-5-yltrimethylstannane. A dark solid was obtained in a yield of 86%. ^{1}H-NMR (300 MHz, CDCl$_3$): δ = 7.80 (d, aromatic H, 2H), 7.70 (d, thiophene-H, 6H), 7.40 (d, thiophene-H, 2H), 7.40 (s, aromatic H, 2H), 6.92 (s, aromatic H, 1H), 6.88 (s, thiophene-H, 1H), 6.60 (d, thiophene-H, 2H), 3.94 (d, CH$_3$-H, 6H). UV/vis (DCM): λ_{max} = 546, 346 nm.

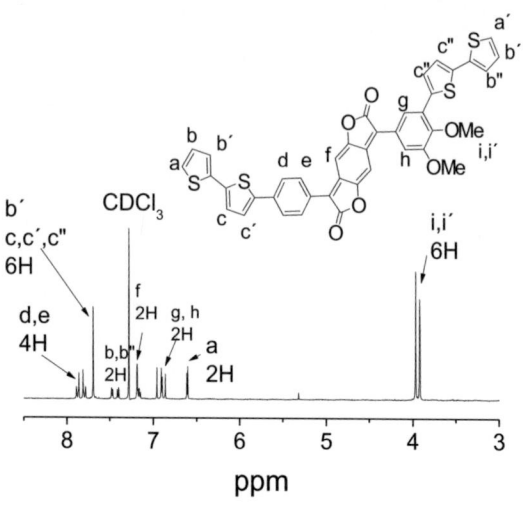

Poly(dimethoxyphenyl-benzodifuranone-dihexylfuorene) (P-uBZDF-FL)

Under N₂, 3-(3-bromo-4,5-dimethoxyphenyl)-7-(4-bromophenyl)benzo[1,2-b:4,5-b']difuran-2,6-dione (**BZDF16**) (150 mg, 0.27 mmol), 2,2'-(9,9-dihexyl-9H-fluorene-2,7-diyl)bis(4,4,5,5-tetramethyl-1,3,2-dioxaborolane) (158 mg, 0.27 mmol), and Pd(PPh$_3$)$_4$ (15 mg, 0.013 mmol) were dissolved in 20 ml toluene. The reaction mixture was degassed and heated to 50 °C. A degassed solution of potassium carbonate (110 mg, 0.81 mml) in 5 ml water was added. The mixture was stirred at 100 °C for 10 h. After cooling, the dark purple solution was diluted with 50 ml DCM, and washed with 50 ml saturated brine and water twice. The organic phase was dried over MgSO$_4$, and the solvent was evaporated. The polymer was dissolved again with a small amount of DCM and precipitated in methanol. (196 mg, 70%). ^{1}H-NMR (300 MHz, CDCl$_3$): δ = 7.80 - 6.20 (m, aromatic H), 4.03 (m, CH3-H), 2.18 - 1.60 (m, α-CH$_2$-H), 1.35 -

200

0.90 (m, CH$_2$-H), 0.86 - 0.51 (m, CH$_3$-H). UV/Vis (DCM): λ_{max} = 532 nm. Fluorescence (DCM): λ_{max} = 711 nm. ε(532) = 28 000 L mol^{-1}cm^{-1}. Molecular weight: 26 000 Da. Polydispersity: 3.5.

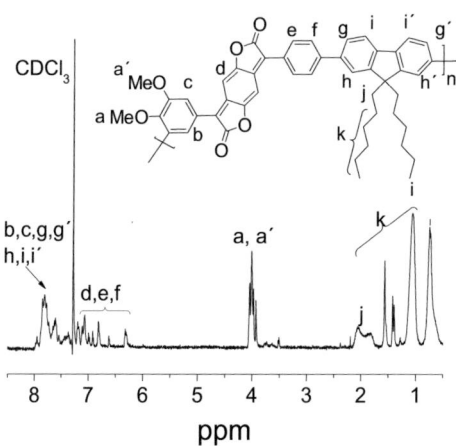

Poly(dimethoxyphenyl-benzodifuranone-thiophene) (P-uBZDF-TH)

Under N$_2$, 3-(3-Bromo-4,5-dimethoxyphenyl)-7-(4-bromophenyl)benzo[1,2-b:4,5-b']difuran-2,6-dione (**BZDF16**) (120 mg, 0.22 mmol), 2,5-bis(trimethylstannyl)thiophene (88 mg, 0.22 mmol), and Pd(PPh$_3$)$_4$ (12 mg, 0.01 mmol) were dissolved in 20 ml dry THF. The reaction mixture was degassed, and refluxed for 10 h under stirring. After cooling, the dark purple solution was diluted with 50 ml DCM, and washed with 50 ml saturated brine and water twice. The organic phase was dried over MgSO$_4$, filtered through a celite layer, and the solvent was evaporated. The polymer was dissolved again with a small account of DCM and precipitated in methanol. (103 mg, 76%). ^{1}H-NMR (300 MHz, CDCl$_3$): δ = 8.01 - 6.44 (m, aromatic H),

3.97 (m, CH$_3$-H). UV/Vis (DCM): λ_{max} = 557, 404 nm. Fluorescence (DCM): λ_{max} = 719 nm. ε(532) = 32 500 L mol^{-1}cm^{-1}. Molecular weight: 9 500 Da. Polydispersity: 2.0.

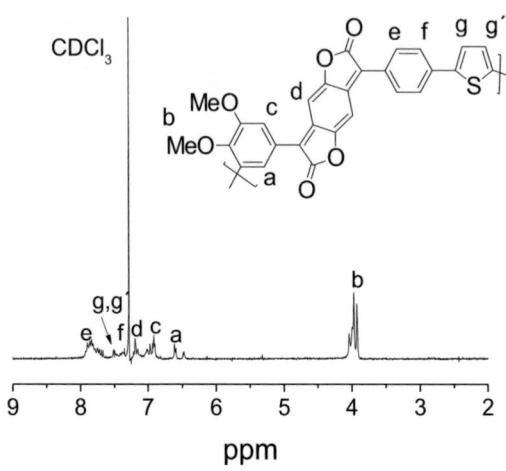

3-(4-Bromophenyl)-7-(3,4-dimethoxyphenyl)benzo[1,2-b:4,5-b']difuran-2,6-dione (BZDF)

The same procedure as described for compound **BZDF16** was used except that 3-bromo-4,5-dimethoxy-mandelic acid was replaced by 4,5-dimethoxy-mandelic acid. The title compound was obtained as a dark brown solid (52%). ^{1}H-NMR (300 MHz, CDCl$_3$): δ = 7.71 (d, aromatic H, 4H), 7.48 (s, aromatic H, 1H), 7.02 (d, aromatic H, 1H), 6.96 (s, aromatic H, 1H), 6.87 8s, aromatic H, 1H), 3.99 (s, CH$_3$-H, 6H).

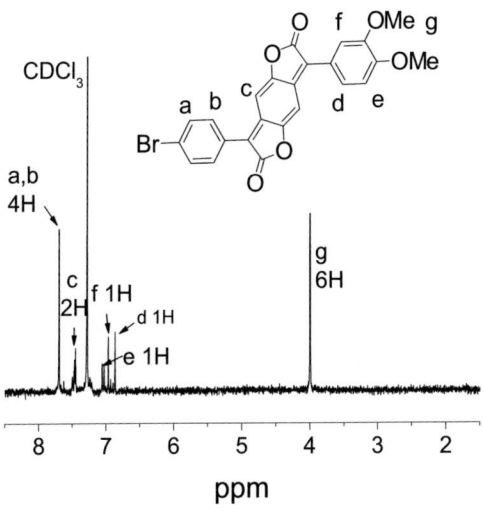

6. References

1. C. K. Chiang, C. R. Fincher, Y. W. Park, A. J. Heeger, H. Shirakawa, E. J. Louis, S. C. Gau and A. G. MacDiarmid, *Phys. Revi. Lett.*, 1977, **39**, 1098.
2. C. K. Chiang, Y. W. Park, A. J. Heeger, H. Shirakawa, E. J. Louis and A. G. Macdiarmid, *J. Chem. Phys.*, 1978, **69**, 5098-5104.
3. H. Shirakawa, E. J. Louis, A. G. MacDiarmid, C. K. Chiang and A. J. Heeger, *Chem . Commun.*, 1977, 578-580.
4. G. Yu, J. Gao, J. C. Hummelen, F. Wudl and A. J. Heeger, *Science*, 1995, **270**, 1789.
5. J. Y. Kim, K. Lee, N. E. Coates, D. Moses, T. Q. Nguyen, M. Dante and A. J. Heeger, *Science*, 2007, **317**, 222.
6. K. M. Coakley and M. D. McGehee, *Chem. Mater.*, 2004, **16**, 4533.
7. J. Brabec, N. S. Sariciftci and J. C. Hummelen, *Adv. Funct. Mater.*, 2001, **11**, 15.
8. C. J. Brabec, N. S. Sariciftci and J. C. Hummelen, *Adv. Funct. Mater.*, 2001, **11**, 15.
9. S. Günes, H. Neugebauer and N. S. Sariciftci, *Chem. Rev.*, 2007, **107**, 1324.
10. J. H. Burroughes, D. D. C. Bradley, A. R. Brown, R. N. Marks, K. Mackay, R. H. Friend, P. L. Burns and A. B. Holmes, *Nature*, 1990, **347**, 539.
11. H. Koezuka, A. Tsumura and T. Ando, *Synth. Met.*, 1987, **18**, 699-704.
12. Z. Hao and A. Iqbal, *Chem. Soc. Rev.*, 1997, **26**, 203.
13. K. Zhang and B. Tieke, *Macromolecules*, 2008, **41**, 7287.
14. D. G. Farnum, G. Metha, G. G. I. Moore and F. P. Siegal, *Tetrahedron Lett.*, 1974, **29**, 2549.
15. A. Iqbal and L. Cassar, *U.S. Patent 4,415,685*, 1983.
16. A. C. Rochat, L. Cassar and A. Iqbal, *EP 94911*, 1983.
17. J. Mizuguchi and G. Rihs, *Ber. Bunsen-Ges. Phys. Chem.*, 1992, **96**, 597.
18. J. Mizuguchi, A. Grubenmann, G. Wooden and G. Rihs, *Acta Crystallogr.*, 1992, **B48**, 696.
19. J. Mizuguchi, *J. Phys. Chem. A* 2000, **104**, 1817.
20. T. Potrawa and H. Langhals, *Chem. Ber.*, 1987, **120**, 1075.
21. H. Langhals, T. Grundel, T. Potrawa and K. Polborn, *Liebigs. Ann.*, 1996, 679.
22. J. S. Zombounis, Z. Hao and A. Iqbal, *Nature* 1977, **388**, 131.
23. W.-K. Chan, Y. Chen, Z. Peng and L. Yu, *J. Am. Chem. Soc.*, 1993, **115**, 11735.
24. S. H. Eldin, A. Iqbal and Z. Hao, *EP 0787730*, 1997.
25. S. Eldin and S. H. Iqbal, *EP 0787731*, 1997.
26. W. K. Chan, Y. Chen, Z. Peng and L. Yu, *J. Am. Chem. Soc.*, 1993, **115**, 11735.
27. G. Lange and B. Tieke, *Macromol. Chem. Phys.*, 1999, **200**, 106.
28. M. Behnke and B. Tieke, *Langmuir*, 2002, **18**, 3815.
29. T. Beyerlein and B. Tieke, *Macromol. Rapid Commun.*, 2000, **21**, 182.
30. M. Smet, B. Mellen and W. Dehaen, *Tetrahedron Lett.*, 2001, 42.
31. M. Horn, Y. Hepuzer, Y. Yagci, B. Bilgin-Eran, U. Cernenco, V. Harabagiu, M. Pinteala and B. C. Simionescu, *Eur. Polym. J.*, 2002, **38**, 2197.

32. T. Beyerlein, B. Tieke, S. Forero-Lenger and W. Brütting, *Synth. Met.*, 2002, **130**, 115.
33. J. Hofkens, W. Verheijen, R. Shukla, W. Dehaen and F. C. De Schryver, *Macromolecules*, 1998, **31**, 4493.
34. A. R. Rabindranath, Y. Zhu, I. Heim and B. Tieke, *Macromolecules*, 2006, **39**, 8250.
35. Y. Zhu, A. R. Rabindranath, T. Beyerlein and B. Tieke, *Macromolecules*, 2007, **40**, 6981.
36. D. Cao, Q. Liu, W. Zeng, S. Han, J. Peng and S. Liu, *J. Polym. Sci., Part A: Polym. Chem.*, 2006, **44**, 2395.
37. A. J. C. Kuehne, A. R. Mackintosh, R. A. Pethrick and B. Tieke, *Tetrahedron Lett.*, 2008, **49**, 4722.
38. Y. Zhu, I. Heim and B. Tieke, *Macromol. Chem. Phys.*, 2006, **207**, 2206.
39. A. R. Rabindranath, Y. Zhu, K. Zhang and B. Tieke, *Polymer*, 2009, **50**, 1637.
40. J. Louie and J. F. Hartwig, *J. Am. Chem. Soc.*, 1997, **119**, 11695.
41. F. E. Doodson and J. F. Hartwig, *Macromolecules*, 1998, **31**, 1700.
42. F. Paul, J. Patt and J. F. Hartwig, *J. Am. Chem. Soc.*, 1994, **116**, 5969-5970.
43. A. S. Guram, R. A. Rennels and S. L. Buchwald, *Angew. Chem., Int. Ed. Engl.*, 1995, **34**, 1348.
44. J.-F. Marcoux, S. Wagaw and S. L. Buchwald, *J. Org. Chem.*, 1997, **62**, 1568.
45. J. P. Wolfe, S. Wagaw and S. L. Buchwald, *J. Am. Chem. Soc.*, 1996, **118**, 7215.
46. T. Kanbara, A. Honma and K. Hasegawa, *Chem. Lett.*, 1996, 1135.
47. T. Kanbara, K. Izumi, Y. Nakadimi, T. Narice and K. Hasegawa, *Chem. Lett.*, 1997, 1185.
48. T. Kanbara, K. Izumi, T. Narice and K. Hasegawa, *Polym. J.*, 1998, **30**, 66.
49. T. Kanbara, M. Oshima and K. Hasegawa, *Macromolecules*, 1998, **31**, 8725.
50. L. Bürgi, M. Turbiez, R. Pfeiffer, F. Bienewald, H.-J. Kirner and C. Winnewisser, *Adv. Mater.*, 2008, **20**, 2217.
51. J. C. Bijleveld, A. Zoombelt, S. G. J. Mathijssen, M. M. Wienk, M. Turbiez, D. M. de Leeuw and R. A. J. Janssen, *J. Am. Chem. Soc.*, 2009, **131**, 16616.
52. Y. P. Zou, D. Gendron, R. Neagu-Plesu and M. Leclerc, *Macromolecules*, 2009, **42**, 6361.
53. Y. P. Zou, D. Gendron, R. Badrou-Aïch, A. Najari, Y. Tao and M. Leclerc, *Macromolecules*, 2009, **42**, 2891.
54. L. Huo, J. Hou, H.-Y. Chen, S. Zhang, Y. Jiang, T. Chen and Y. Yang, *Macromolecules*, 2009, **42**, 6504.
55. E. Zhou, Q. Wei, S. Yamakawa, Y. Zhang, K. Tajima, C. Yang and K. Hashimoto, *Macromolecules*, 2010, **43**, 821.
56. E. Zhou, S. Yamakawa, K. Tajima, C. Yang and K. Hashimoto, *Chem. Mater.*, 2009, **21**, 4055.
57. C. Kanimozhi, D. Baljaru, G. D. Sharma and S. Patil, *J. Phys. Chem. B*, 2010, **114**, 3095.
58. Y. Zou, D. Gendron, R. Badrou-Aïch, A. Najari, Y. Tao and M. Leclerc, *Macromolecules*, 2009, **42**, 2891.
59. C. Kanimozhi, D. Baljaru, G. D. Sharma and S. Patil, *J. Phys. Chem. C*, 2010, **114**, 3287.
60. A. P. Zoombelt, S. G. J. Mathijssen, M. G. R. Turbiez, M. M. Wienk and R. A. J. Janssen, *J. Mater. Chem.*, 2010, **20**, 2240.
61. H. Junek, *Monatch. Chem.*, 1960, **91**, 479.
62. C. W. Greenhalgh, J. L. Carey and D. F. Newton, *Dyes Pigm.*, 1980, **1**, 103.

63. C. W. Greenhalgh, J. L. Carey, N. Hall and D. F. Newton, *J. Soc. Dyes Colour.*, 1994, **110**, 178.
64. O. Annen, R. Egli, B. Henzi, H. Jacob and P. Matzinger, *Rev. Prog. Coloration*, 1987, **17**, 72.
65. A. A. Gorman, M. G. Huutchings and P. H. Wood, *J. Am. Chem. Soc.*, 1996, **118**, 8497.
66. G. Hallas and C. Yoon, *Dyes Pigm.*, 2001, **48**, 107.
67. T. Yamamoto, A. Morita, Y. Miyazaki, T. Maruyama, H. Wakayama, Z.-H. Zhou, Y. Nakamura and T. Kanbara, *Macromolecules*, 1992, **25**, 1214.
68. S. Saito and Y. Yamamoto, *Chem. Rev.*, 2000, **100**, 2901.
69. E. Negishi, 1998.
70. R. K. Heck, *Org. React.*, 1982, **27**, 345.
71. N. Miyaura and A. Suzuki, *Chem. Rev.*, 1995, **95**, 2457-2483.
72. J. K. Stille, *Angew. Chem. Int. Edit.*, 1986, **25**, 508-523.
73. D. R. Rutherford, J. K. Stille, C. M. Elliott and V. R. Reichert, *Macromolecules*, 1992, **25**, 2294.
74. H. Masai, K. Sonogashira and N. Hagihara, *Bull. Chem. Soc. Jpn.*, 1971, **44**, 2226.
75. K. Sonogashira, Y. Tohda and N. Hagihara, *Tetrahedron Lett.*, 1975, **16**, 4467-4470.
76. K. Sonogashira, *J. Organomet. Chem.*, 2002, **653**, 46.
77. A. S. Guram and S. L. Buchwald, *J. Am. Chem. Soc.*, 1994, **16**, 7901-7902.
78. P. E. Fanta and F. Ullmann, *Chem. Rev.*, 1964, **64**, 613.
79. T. Gibtner, F. Hampel, J. P. Gisselbrecht and A. Hirsch, *Chem. Eur. J.*, 2002, **8**, 408-432.
80. R. Chinchilla and C. Nájera, *Chem. Rev.*, 2007, **107**, 874-922.
81. Z. Adamcova and L. Dempirova, *Prog. Org. Coat.*, 1989, **16**, 295-320.
82. M. Zhou and J. Heinze, *Electrochim. Acta*, 1999, **44**, 1733.
83. M. Zhou and J. Heinze, *J. Phys. Chem. B*, 1999, **103**, 8443.
84. M. Zhou and J. Heinze, *J. Phys. Chem. B*, 1999, **103**, 8451.
85. M. Zhou, V. Rang and J. Heinze, *Acta Chem. Scand.*, 1999, **53**, 1059.
86. M. Zhou, M. Pagels, B. Geschke and J. Heinze, *J. Phys. Chem. B*, 2002, **106**, 10065-10073.
87. S. Sadki, P. Schottland, N. Brodie and G. Sabouraud, *Chem. Soc. Rev.*, 2000, **29**, 283.
88. A. F. Diaz, J. L. Castillo, J. A. Logan and W. Y. Lee, *J. Electroanal. Chem.*, 1982, 118.
89. C. A. Thomas, K. Zong, P. Schottland and J. R. Reynolds, *Adv. Mater.*, 2000, **12**, 222.
90. C. L. Gaupp, K. Zong, P. Schottland, B. C. Thompson and J. R. Reynolds, *Macromolecules*, 2000, **33**, 1132.
91. C. L. Gaupp, D. M. Welsh and J. R. Reynolds, *Macromol. Rapid Commun.*, 2002, **23**, 885.
92. G. A. Sotzing, J. R. Reynolds and P. Steel, *J. Chem. Mater.*, 1996, **8**, 882.
93. G. A. Sotzing, J. L. Reddinger, A. R. Katritzky, J. Soloducho, R. Musgrave and J. R. Reynolds, *Chem. Mater.*, 1997, **9**, 1578.
94. S. X. Deng and R. C. Advincula, *Chem. Mater.*, 2002, **14**, 4073-4080.
95. M. Kabasakaloglu, T. Kiyak, H. Toprak and M. L. Aksu, *Appl. Surf. Sci.*, 1999, **152**, 115-125.
96. G. Q. Shi, C. Li and Y. Q. Liang, *Adv. Mater.*, 1999, **11**, 1145.
97. A. D. Child, B. Sankaran, F. Larmat and J. R. Reynolds, *Macromolecules*, 1995, **28**, 6571-6578.
98. C. S. Lin, J. A. Chen, M. H. Liu, Y. O. Su and M. Leung, *J. Org. Chem.*, 1998, **63**, 149-156.

99. S. Destri, G. U., A. Fazio, W. Porzio, B. Gabriele and G. Zotti, *Org. Electron.*, 2002, **3**, 149-156.
100. C.-C. You, C. R. Saha-Moller and F. Wurthner, *Chem. Commun.*, 2004, 2030-2031.
101. G. Friedrich, H. Gerhard, S. Werner, H. Jurgen and D. Michael, *EP 0339340*, 1989
102. G. Heywang and F. Jonas, *Adv. Mater.*, 1992, **4**, 116-118.
103. F. Jonas and L. Schrader, *Synth. Met.*, 1991, **41**, 831-836.
104. G. Zotti, S. Zecchin, G. Schiavon and L. Groenendaal, *Macromol. Chem. Phys.*, 2002, **201**, 1958-1964.
105. A. Czardybon and M. Lapkowski, *Synth. Met.*, 2001, **119**, 161-162.
106. M. Turbiez, P. Frère, M. Allain, N. Gallego-Planas and J. Roncali, *Macromolecules*, 2005, **38**, 6806.
107. H. W. Heuer, R. Wehrmann and S. Kirschmeyer, *Adv. Funct. Mater.*, 2002, **12**, 89-94.
108. I. Schwendeman, R. Hickman, G. Sonmez, P. Schottland, K. Zong, D. W. Welsh and J. R. Reynolds, *Chem. Mater.*, 2002, **14**, 3118.
109. C. Wang, J. L. Schindler, C. R. Kannewurf and M. G. Kanatzidis, *Chem. Mater.*, 1995, **7**, 58-68.
110. F. Goldoni, B. M. W. Langeveld-Voss and E. W. Meijer, *Synth. Commun.*, 1998, **28**, 2237.
111. H. J. Spencer, P. J. Skabara, M. Giles, I. McCulloch, S. J. Coles and M. B. Hursthouse, *J. Mater. Chem.*, 2005, **15**, 4783-4792.
112. H. Taoudi, J. C. Bernède, M. A. Del Valle, A. Bonnet and M. Morsli, *J. Mater. Sci.*, 2001, **36**, 631-634.
113. M. Li, S. Tang, F. Shen, M. Liu, W. Xie, H. Xia, L. Liu, L. Tian, Z. Xie, P. Lu, M. Hanif, D. Lu, G. Cheng and Y. Ma, *J. Phys. Chem. B*, 2006, **110**, 17784-17789.
114. M.-Y. Chou, M.-k. Leung, Y. O. Su, C. L. Chiang, C.-C. Lin, J.-H. Liu, C.-K. Kuo and C.-Y. Mou, *Chem. Mater.*, 2004, **16**, 654-661.
115. N. S. Sariciftci, L. Smilowitz, A. J. Heeger and F. Wudl, *Science*, 1992, **258**, 1474.
116. B. R. Weinberger, M. Akhtar and S. C. Gau, *Synth. Metals*, 1982, **44**, 187-197.
117. S. Glenis and G. Tourillon, *Thin Solid Films*, 1984, **139**, 221-231.
118. C. W. Tang, *Appl. Phys. Lett.*, 1986, **48**, 183.
119. H. Hoppe and N. S. Sariciftci, *J. Mater. Chem.*, 2004, **19**, 1924.
120. J. J. M. Halls, K. Pichler, R. H. Friend, S. C. Moratti and A. B. Holmes, *Appl. Phys. Lett.*, 1996, **68**, 3120.
121. M. Theander, A. Yartsev, D. Zigmantas, V. Sundström, W. Mammo, M. R. Andersson and O. Inganäs, *Phys. Rev. B*, 2000, **61**, 12957.
122. A. Haugeneder, M. Neges, C. Kallinger, W. Spirkl, U. Lemmer, J. Feldmann, U. Scherf, E. Harth, A. Gügel and K. Müllen, *Phys. Rev. B*, 1999, **59**, 15346.
123. T. Stübinger and W. Brütting, *J. Appl. Phys.*, 2001, **90**, 3632.
124. D. E. Markov, E. Amsterdam, P. W. M. Blom, A. B. Sieval and J. C. Hummelen, *J. Phys. Chem. A*, 2005, **109**, 5266.
125. J. C. Hummelen, B. W. Knight, F. LePeg and F. Wudl, *J. Org. Chem.*, 1995, **60**, 532.
126. M. Wienk, J. M. Kroon, W. J. H. Verhees, J. C. Hummelen, P. A. Van Hal and J. Janssen, *Angew. Chem., Int. Ed.*, 2003, **42**, 3371.

127. Y. Yao, C. Shi, G. Li, V. Shrotriya, Z. Pei and Y. Yang, *Appl. Phys. Lett.*, 2006, **89**, 153507.
128. C. Brabec, C. Winder, N. S. Sariciftci, J. C. Hummelen, A. Dhanabalan, P. A. van Hal and R. A. Janssen, *Adv. Funct. Mater.*, 2002, **12**, 709.
129. E. Bundgaard and F. C. Krebs, *Sol. Energy Mater. Sol. Cells*, 2007, **91**, 954.
130. B. C. Thompson and J. M. J. Fréchet, *Angew. Chem., Int. Ed.*, 2008, **47**, 58.
131. A. C. Mayer, S. R. Scully, B. E. Hardin, M. W. Rowell and M. D. McGehee, *Mater. Today*, 2007, **10**, 28.
132. Y.-J. Cheng, S.-H. Yang and C.-S. Hsu, *Chem. Rev.*, 2009, **109**, 5868-5923.
133. R. C. Coffin, J. Peet, J. Rogers and G. C. Bazan, *Nat. Chem.*, 2009, **1**, 657-661.
134. G. Zhao, Y. He and Y. Li, *Adv. Mater.*, 2010, 4355-4358.
135. A. Bernanose, M. Comte and P. Vouaux, *J. Chim. Phys.*, 1953, **50**, 64.
136. A. Bernanose, *J. Chim. Phys.*, 1955, **52**, 396.
137. A. Bernanose and P. Vouaux, *J. Chim. Phys.*, 1955, **52**, 509.
138. C. W. Tang and S. A. VanSlyke, *Appl. Phys. Lett.*, 1987, **51**, 913-915.
139. J. H. Burroughes, D. D. C. Bradley, A. R. Brown, R. N. Marks, K. Mackay, R. H. Friend, P. L. Burns and A. B. Holmes, *Nature*, 1990, **347**, 539-541.
140. R. Mortimer, *J. Electrochim. Acta*, 1999, **44**, 2971.
141. E. Unur, P. M. Beaujuge, S. Ellinger, J.-H. Jung and J. R. Reynolds, *Chem. Mater.*, 2009, **21**, 5145-5153.
142. P. M. Budd, B. S. Ghanem, S. Makhseed, N. B. McKeown, K. J. Msayib and C. E. Tattershall, *Chem. Commun.*, 2004, 230–231.
143. P. M. Budd, A. Butler, J. Selbie, K. Mahmood, N. B. McKeown, B. Ghanem, K. Msayib, D. Book and A. Walton, *Phys. Chem. Chem. Phys.*, 2007, **9**, 1802-1808.
144. N. B. McKeown, B. M. Gahnem, K. Msayib, P. M. Budd, C. E. Tattershall, K. Mahmood, S. Tan, D. Book, H. W. Langmi and A. Walton, *Angew. Chem., Int. Ed.*, 2006, **45**, 1804-1807.
145. B. S. Ghanem, K. J. Msayib, N. B. McKeown, K. D. M. Harris, Z. Pan, P. M. Budd, A. Butler, J. Selbie, D. Book and A. Walton, *Chem. Commun.*, 2007, 67-69.
146. M. P. Tsyurupa and V. A. Davankov, *React. Funct. Polym.*, 2006, **66**, 768.
147. J. Y. Lee, C. D. Wood, D. Bradshaw, M. J. Rosseinsky and A. I. Cooper, *Chem. Commun.*, 2006, 2670.
148. C. D. Wood, B. Tan, A. Trewin, H. J. Niu, D. Bradshaw, M. J. Rosseinsky, Y. Z. Khimyak, N. L. Campbell, R. Kirk, E. Stockel and A. I. Cooper, *Chem. Mater.*, 2007, **19**, 2034.
149. A. P. Côté, A. I. Benin, N. W. Ockwig, M. O'Keeffe, A. J. Matzger and O. M. Yaghi, *Science*, 2005, **310**, 1166.
150. H. M. El-Kaderi, J. R. Hunt, J. L. Mendoza-Corte´s, A. P. Côté, R. E. Taylor, M. O'Keeffe and O. M. Yaghi, *Science*, 2007, **316**, 268.
151. N. B. McKeown, P. M. Budd, K. J. Msayib, B. S. Ghanem, H. J. Kingston, C. E. Tattershall, S. Makhseed, K. J. Reynolds and D. Fritsch, *Chem. Eur. J.*, 2005, **11**, 2610.
152. C. D. Wood, B. Tan, A. Trewin, F. Su, M. J. Rosseinsky, D. Bradshaw, Y. Sun, L. Zhou and A. I. Cooper, *Adv. Mater.*, 2008, **20**, 1916.
153. H. J. Mackintosh, P. M. Budd and N. B. McKeown, *J. Mater. Chem.*, 2008, **18**, 537.

154. H. B. Park, C. H. Jung, Y. M. Lee, A. J. Hill, S. J. Pas, S. T. Mudie, E. V. Wagner, B. D. Freeman and D. J. Cookson, *Science*, 2007, **318**, 254.
155. S. Brunauer, P. H. Emmett and E. Teller, *J. Am. Chem. Soc.*, 1938, **60**, 309.
156. J. H. Ahn, J. E. Jang, C. G. Oh, S. K. Ihm, J. Cortez and D. C. Sherrington, *Macromolecules*, 2006, **39**, 627.
157. J.-X. Jiang, A. Trewin, F. Su, C. D. Wood, H. Niu, J. T. A. Jones, Y. Z. Khimyak and A. I. Cooper, *Macromolecules*, 2009, **42**, 2658-2666.
158. A. I. Cooper, *Adv. Mater.*, 2009, **21**, 1292-1295.
159. J.-X. Jiang, F. Su, A. Trewin, C. D. Wood, H. Niu, J. T. A. Jones, Y. Z. Khimyak and A. I. Cooper, *J. Am. Chem. Soc.*, 2008, **130**, 7710-7720.
160. N. Kobayashi and M. Kijima, *J. Mater. Chem.*, 2007, **17**, 4289.
161. J. X. Jiang, F. Su, A. Trewin, C. D. Wood, N. L. Campbell, H. Niu, C. Dickinson, A. Y. Ganin, M. J. Rosseinsky, Y. Z. Khimyak and A. I. Cooper, *Angew. Chem., Int. Ed.*, 2007, **46**, 8574.
162. J. X. Jiang, F. Su, H. J. Niu, C. D. Wood, N. L. Campbell, Y. Z. Khimyak and A. I. Cooper, *Chem. Commun.*, 2008, 486-488.
163. P. Kuhn, M. Antonietti and A. Thomas, *Angew. Chem. Int. Ed.*, 2008, **47**, 3450.
164. I. Langmuir, *J. Am. Chem. Soc.*, 1916, **38**, 2221-2295.
165. P. W. Atkins, *Physical Chemistry*, W.H. Freeman & Company, 1997.
166. E. P. Barrett, L. G. Joyner and P. P. Halenda, *J. Am. Chem. Soc.*, 1951, **73**, 373-380.
167. B. C. Lippens, B. G. Linsen and J. H. D. Boer, *J. Catal.*, 1964, **3**, 32-37.
168. J. H. D. Boer, B. C. Lippens, B. G. Linsen, J. C. P. Broekhoff, A. van den Heuvel and T. J. Osinga, *J. Colloid Interface Sci.*, 1966, **21**, 405-414.
169. B. C. Lippens and J. H. de Boer, *J. Catal.*, 1965, **4**, 319.
170. W. Harkins and G. Jura, *J. A. Chem. Soc.*, 1944, **66**, 1362.
171. G. D. Halsey, *J. Chem. Phys.*, 1948, **16**, 31.
172. J. Hwang, D. B. Tanner, I. Schwendenman and J. R. Reynolds, *Phys. Rev. B* 2003, **67**, 115105.
173. H. J. Spencer, P. J. Skabara, M. Giles, I. McCulloch, S. J. Coles and M. B. Hursthouse, *J. Mater. Chem.*, 2005, **15**, 4783.
174. D. J. Crouch, P. J. Skabara, J. E. Lohr, J. W. McDouall, M. Heeney, I. McCulloch, D. Sparrowe, M. Shkunov, S. J. Coles, P. N. Horton and M. B. Hursthouse, *Chem. Mater.*, 2005, **17**, 6567.
175. R. Berridge, S. P. Wright, P. J. Skabara, A. Dyer, T. Steckler, A. A. Argun, J. R. Reynolds, R. W. Harrington and W. Clegg, *J. Mater. Chem.*, 2007, **17**, 225.
176. Y. Zhu, K. Zhang and B. Tieke, *Macromol. Chem. Phys.*, 2009, **210**, 431.
177. J. L. Bredas, R. Silbey, D. S. Boudreux and R. R. Chance, *J. Am. Chem. Soc.*, 1983, **105**, 6555.
178. L. Groenendaal, F. Jonas, D. Freitag, H. Pielartzik and J. R. Reynolds, *Adv. Mater.*, 2000, **12**, 48.
179. J. Roncali, R. Garreau, A. Yassar, P. Marque, F. Garnier and M. Lemaire, *J. Phys. Chem.*, 1987, **91**, 6706.

180. K. Zhang, B. Tieke, J. C. Forgie and P. J. Skabara, *Macromol. Rapid Commun.*, 2009, **30**, 1834.
181. H. J. Spencer, P. J. Skabara, M. Giles, I. McCulloch, S. J. Coles and M. Hursthouse, *J. Mater. Chem.*, 2005, **15**, 4793.
182. P. J. Skabara, R. Berridge, E. J. L. McInnes, D. P. West, S. J. Coles, M. B. Hursthouse and K. Müllen, *J. Mater. Chem.*, 2004, **14**, 1964.
183. E. P. Barrett, L. G. Joyner and P. P. Halenda, *J. Am. Chem. Soc.*, 1951, **73**, 373.
184. J. S. Zambounis, Z. Hao and A. Iqbal, *Nature*, 1997, **388**, 131.
185. P. J. Skabara, *Fused Oligothiophenes, Chapter in Handbook of Thiophene-Based Materials* John Wiley & Sons, 2009.
186. P. T. Henderson and D. M. Collard, *Chem. Mater.*, 1995, **7**, 1879.
187. D. D. Perrin and W. L. F. Armarego, *Purification of Laboratory Chemicals*, Pergamon, 1988.
188. A. A. Argun, P.-H. Aubert, B. C. Thompson, I. Schwendeman, C. L. Gaupp, J. Hwang, N. J. Pinto, D. B. Tanner, A. G. MacDiarmid and J. R. Reynolds, *Chem. Mater.*, 2004, **16**, 4401-4412.
189. J. Hou, M. H. Park, S. Zhang, Y. Yao, L. M. Chen, J. H. Li and Y. Yang, *Macromolecules*, 2008, **41**, 6012.
190. B. E. Maryanoff, D. F. McComsey, J. F. Gardocki, R. P. Shank, M. J. Costanzo, S. O. Nortey, C. R. Schneider and P. E. Setler, *J. Med. Chem.*, 1987, **30**, 1433-1454.
191. H. Shapiro, K. A. Smith, D. H. S. Horn and F. L. Warren, *J. Chem. Soc.*, 1946, 143-144.

Publications

Part of Publications:

Zhang, K.; Tieke, B.; Forgie, J. C.; Skabara, P. J., "Electrochemical Polymerization of N-Arylated and N-Alkylated EDOT-Substituted Pyrrolo[3,4-c]pyrrole-1,4-dione (DPP) Derivatives: Influence of Substitution Pattern on Optical and Electronic Properties", *Macromol. Rapid Commun.* **2009**, 30, 1834.

Tieke, B.; Rabindranath, A. R.; Zhang, K.; Zhu, Y., "Conjugated polymers containing diketopyrrolopyrrole units in the main chain", *Beilstein J. Org. Chem.* **2010**, 6, 830.

Zhang, K.; Tieke. B.; Forgie, J. C.; Vilela, F.; Skabara, P. J. *Polymer*, "Cross-linked polymers based on 2,3,5,6-tetra-substituted pyrrolo[3,4-c]pyrrole-1,4(2H,5H)-dione (DPP): Synthesis, optical and electronic properties", *Polymer*, **2010**, 51, 26, 6107.

Kanibolotsky, A. L.; Vilela, F.; Forgie, J. C.; Elmasly, S.; Skabara, P. J.; Zhang, K.; Tieke, B.; McGurk, J.; Belton, C. R.; Stavrinou, P. N.; Bradley, D. D. C., "Well-defined and monodisperse linear and star-shaped quaterfluorene-DPP molecules: the significance of conjugation and dimensionality", *Adv. Mater.* **2011**, 23, 2093.

Zhang, K.; Tieke, B.; Vilela, F.; Skabara, P. J., "Conjugated Microporous Networks on the Basis of 2,3,5,6-Tetraarylated Diketopyrrolo[3,4-c]pyrrole", *Macromol. Rapid Commun.* **2011**, 32, 825.

Zhang, K.; Tieke, B.,"Low-Bandgap Benzodifuranone-Based Polymers", *Macromolecules* **2011**, 44, 4596.

Zhang, K.; Tieke, B.; Forgie, J. C.; Vilela, F.; Skabara, P. J.,"Donor-Acceptor Conjugated Polymers Based on *p*- and *o*-Benzodifuranone and Thiophene Derivatives: Electrochemical Preparation and Optical and Electronic Properties", *Macromolecules* **2012**, 45, 743.

Der disserta Verlag bietet die kostenlose Publikation
Ihrer Dissertation als hochwertige
Hardcover- oder Paperback-Ausgabe.

Fachautoren bietet der disserta Verlag
die kostenlose Veröffentlichung professioneller Fachbücher.

Der disserta Verlag ist Partner für die Veröffentlichung
von Schriftenreihen aus Hochschule und Wissenschaft.

Weitere Informationen auf www.disserta-verlag.de